D1372703

A SWAMP FULL OF DOLLARS

A SWAMP FULL OF DOLLARS

PIPELINES AND PARAMILITARIES AT NIGERIA'S OIL FRONTIER

MICHAEL PEEL

Lawrence Hill Books

This edition published 2010 by Lawrence Hill Books
First published by I.B. Tauris & Co Ltd in the United Kingdom
© 2009 by Michael Peel
All rights reserved

Published by Lawrence Hill Books
An imprint of Chicago Review Press, Incorporated
814 North Franklin Street
Chicago, Illinois 60610
ISBN 978-1-56976-286-8
Printed in the United States of America
5 4 3 2 1

To my family and in particular my parents,
Mary and Robin Peel

CONTENTS

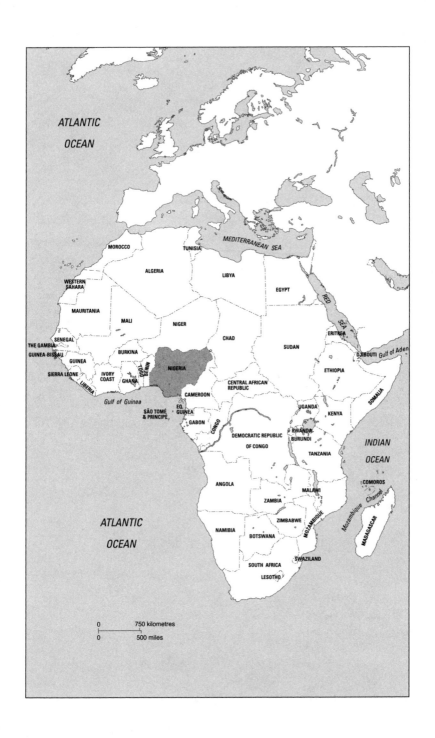

ATLANTIC
OCEAN

MEDITERRANEAN SEA

MOROCCO

TUNISIA

WESTERN
SAHARA

ALGERIA

LIBYA

EGYPT

RED SEA

MAURITANIA

MALI

NIGER

CHAD

SUDAN

ERITREA

DJIBOUTI Gulf of Aden

THE GAMBIA
SENEGAL
GUINEA-BISSAU
GUINEA
SIERRA LEONE
LIBERIA

BURKINA

IVORY
COAST

GHANA

TOGO
BENIN

NIGERIA

CAMEROON

CENTRAL AFRICAN
REPUBLIC

ETHIOPIA

SOMALIA

Gulf of Guinea

SÃO TOMÉ
& PRÍNCIPE

EQ.
GUINEA

GABON

CONGO

DEMOCRATIC REPUBLIC
OF CONGO

UGANDA

RWANDA
BURUNDI

KENYA

TANZANIA

INDIAN
OCEAN

COMOROS

ATLANTIC
OCEAN

ANGOLA

ZAMBIA

ZIMBABWE

MALAWI

MOZAMBIQUE

Mozambique Channel

MADAGASCAR

NAMIBIA

BOTSWANA

SOUTH AFRICA

SWAZILAND

LESOTHO

0   750 kilometres
0   500 miles

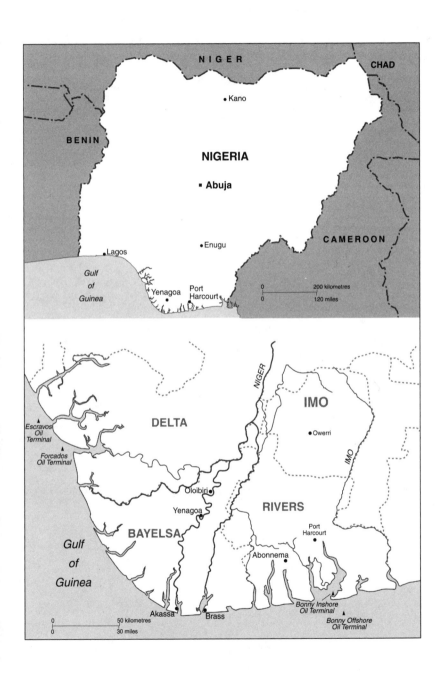

PROLOGUE:
TRIGGER POINT

It is late, almost too late, to be looking for oil. The thought grows in my mind as we creep and crunch up the gravel road to the old cocoa plantation at Uba Budu, which seems as lofty and remote as an Alpine ski station. My driver is skilled, but understandably cautious: it is, after all, his car's chassis that will be mangled by any stray rocks. It is deep into the afternoon now; by six, the forest will be dark and its landmarks amorphous. I won't be able to see the *agua petróleo*, the crude-filled pool rumoured to lie at the heart of this West African jungle. My hunt for black gold will have run out of time.

We are on the island of São Tomé and Príncipe, a short hop across the Gulf of Guinea coast from Nigeria's oil-rich Niger Delta. São Tomé is reputed to be blessed – or cursed – with crude oil too, although it's hard to imagine in the magisterial stillness of this virgin forest. We pass little that is man-made, save the crumbling old walls built by Portuguese colonialists who thought they would rule forever but ended up leaving in a hurry. We give a lift to an old couple we have seen walking slowly ahead of us, the distance they keep between them suggesting long familiarity. Both their faces are etched with deep lines, like the tyre treads we are following up the forest track. The man is taciturn but the woman talks urgently in creole, communicating a controlled desperation through the remainder of the drive.

It is almost four by the time we reach the first of Uba Budu's ghostly houses, which appear as if frozen at the instant they were abandoned more than three decades ago. In the plantation's main square, a group of young men watches us from a small outhouse set a little apart from the sprawling white main building. The old woman gets out of the car, looking at me and raising her hand to her mouth in a plea for money. When I give her dobras worth about £1.50, she grips my hand with a strength unnerving in one apparently so frail. The intensity of her gratitude fills me with loathing, both for the economic gulf from which it springs and for the feeling of power it awakens in me.

A few of the men wander over from the outhouse to talk to us, one of them rocking along on crutches several yards behind the main group. The oldest of them, his hair unusually unkempt for a society notable for its high standards of personal grooming, wags his finger in warning when he hears our plan. He doesn't rate our chances of reaching the *agua petróleo* before nightfall, although he is reluctant to tell us exactly how far away it is. His negativity makes me cussed and determined to go, even though instincts honed in similar situations suggest I should hesitate. The local advice is to hold back, darkness is near, and I am the centre of attention of a crowd of young men. But I remind myself that I am the right side of the frontier of the incendiary Niger Delta, big brother in a partnership between Nigeria and São Tomé to exploit oil. Besides, I now crave sight of the crude reputed to be sitting in the forest, tempting and untouched.

After a few minutes of cajoling, a snake-thin young man named Afocinho Viera agrees to lead us to the *agua*. We start at a fast walk that soon turns into a jog. It's hard work, but comfortable enough, until Viera and Kaizer Montero, another youth who has joined us, dart off the path and into the trees. Soon they are springing down the rocky, loose-soiled paths, moving like mountain goats in flip-flops as I plod like an elephant in my trainers. As we hurtle deeper into a forest where tarantulas and spitting cobras live, I feel elated by the recklessness of the chase. We are alone with the birds and whatever else lurks in the darkening jungle.

Three-quarters of an hour later, my body coated in a slick layer of sweat, I am starting to regret the stubbornness of my pursuit. I feel like a spoilt kid each time I ask Viera and Montero whether we are nearly there. I realize I must cut a ridiculous figure, my shirt flapping open in the breeze like a mad professor's lab coat. Then Viera stops suddenly. He gestures to the right of the stony track, towards a small pool of water flanked by a scattering of giant fallen leaves. Peering down, I see the water has a rusty brown colour and the unmistakable iridescent sheen of oil. It is the *agua petróleo*.

We jump down onto a spit of land, into which my shoes sink as if it were quicksand. Viera leans down by the edge of the pool, which is bubbling gently like a cauldron coming to the boil. He scoops up some liquid and holds his fingers up to me. The brown fluid coating them has the same smell and touch as the light, sweet crude that I've handled in oil spills in the Niger Delta. I cup some myself, gazing with the exaggerated fascination of a small child as it trickles down my fingers, mingling with the perspiration from the effort of reaching this place. I wonder if this is how Africa's resource-hungry foreign adventurers have felt over the centuries, at the moment of grasping the prizes they have bought and fought for so ruthlessly.

Tracing the veins of precious fluid now running in delta-like rivulets across the palm of my hand, I am silent in thought for a moment. After more than seven years of living in and visiting Nigeria and its neighbours, I feel I am finally a step ahead of the expansion of the oil industry whose tumultuous wake I have constantly crossed. This potential oilfield is unobtrusive, unfenced and all but untouched, although I wonder how long this will remain the case. If the Niger Delta is anything to go by, it may be only a matter of time before geological accident thrusts the *agua petróleo* into one of the great geopolitical fights of our age.

One day, I reflect, this hidden pool and its hinterland may look like the warped Niger Delta wonderland of crude that has come to inform the way I see the world. I find it hard now to look at anything connected with oil – whether in the rawness of a Delta slick or the neatness of a Royal Dutch Shell filling station forecourt – without thinking of how it has moulded Nigeria, Africa's leviathan. For me, the logos of Shell, Chevron and ExxonMobil evoke nightmarish images of the great pluming orange gas flares that cast a sickly nocturnal glow over rundown villages, where people drink from stagnant pools not unlike the *agua*. The barrels of oil whose price underpins City economic forecasts are what armed young Delta militants – determined, deluded or drunk, or all three at once – siphon from pipelines to fund a black market stretching across continents. Even in this era of realpolitik, it is hard to imagine a dirtier business in which so many of us in the rich world are so intimately involved.

It's not uncommon for visitors to Nigeria to see it as a disturbing and alien place, its value system warped by oil into something unrecognisable. But, over the years of my association with this brilliant, fragmented nation, I have found my view of its troubles shifting kaleidoscopically. The harder I look, the more I see in the shards a reflection of the life and times of my own homeland, Britain. My story of Nigeria is of a deep and deepening interconnectedness, forged in large part by crude and the wealth and power that flow from it. Like no other place I have visited, Nigeria brings to rich and raucous life the geopolitics of oil that enmesh us all. Its growing problems are ones we share.

Oil-based industries – first palm, then crude – have dominated Nigeria's economy and international relations for more than a century. The hand of Britain can be seen more or less obviously in the colonial era and the unending oil war that threads through the Niger Delta from Victorian times to the present day. But the Anglo-Saxon presence is also there in less expected places, like the apparently anarchic streets of Lagos, a chancers' paradise to rival the City of London and the wider international finance industry. Nigerian politicians are fond of Britain, too: they put their stolen millions through banks there that share their laissez-faire sensibility. If Westerners want to gaze, Scrooge-like, at the disturbing spirit of their age, they need only look to Nigeria to see it expressed more vividly than they might wish.

In the half-century since it shipped its first oil, the nation of Nigeria – one of the world's ten most populous – has become a little laboratory for the arrogance of a fossil-fuel-obsessed world. It is a country where the oil economy is slowly being destroyed by its own hand and its own hubris. It has metamorphosed from pillar of unprecedented Western industrial prosperity to ominous parable for a rapacious age. Nigeria teaches us that the unfettered global cult of crude hurts not just the countries that produce it but – as relentlessly as in any Greek tragedy – the nations that consume it, too. That vulnerability is clear in the world oil market spikes caused by each production disruption in the Delta.

All this makes Nigeria a brittle motor of twenty-first-century capitalism. One of the grimly poetic qualities to what is happening there is that the supply of crude is being undermined partly by the very products and technological developments oil wealth has helped to deliver. The more widely better communications, the possibility of travel and ostentatious consumer goods have become available in the Niger Delta, the more the villagers who live above the oilfields have noticed the difference between their circumstances and those of the people who exploit their natural riches. As the venality and corruption that envelop the industry become clear for all to see, it is hardly surprising that the local battles for Nigerian crude have become increasingly gangsterish. It is, after all, the tried and trusted method to get rich off the country's main resource.

As the world oil price climbed steeply between 2004 and 2008, so the battles over Nigeria's crude became more violent, the Niger Delta turning ever closer to a Mad Max world of roving bandits. Yet, para-doxically – and worryingly – Nigeria and the broader West African region were at the same time assuming an increasingly important role in the energy security policies of Washington and Western allies such as Britain, who are keen for a bulwark against troubles in the Arabian Gulf. Nigeria has historically sent about half its oil production – between 2m and 2.6m barrels of oil daily – to the USA, where it has accounted for about 10 per cent of total imports. The US National Intelligence Council has estimated that African countries, led by Nigeria and Angola, could supply a quarter of its total oil imports by 2015. Nigerian crude is particularly prized, because its low level of impurities makes it ideally suited for refining into gasoline.

The craving for Nigeria's oil has grown also outside the West, reflecting shifts in the global economic balance of power. China is capturing exploration contracts and promising investment in infrastructure projects. In September 2008, Russia's Gazprom signed an agreement to form a joint venture with the state Nigerian National Petroleum Corporation; within a fortnight, the European Union, anxious to reduce its dependence on Russian gas that can be cut at a moment's notice, offered Nigeria financial and political support for a €15bn (£13.3bn) pipeline to send its gas across the Sahara to Europe. If the Arab oil states are the eastern battle-line in the world war for oil, it seems Nigeria is quietly turning into the western front.

My personal engagement with the dark story of the industry that has become almost Nigeria's raison d'être stretches back to my first visit to the country in the months before the 11 September 2001 terrorist attacks. Then, I was taking a break from a deeply glamorous job as the *Financial Times* tax and accountancy correspondent to enter an environment in which everything seemed, as a Nigerian civil war veteran once put it to me, 'too big, too wild and too many'. Almost from the moment I arrived in a Lagos liberally populated by four-by-four off-road vehicles, Mercedes-Benzes and BMWs, I started to wonder about the fate of the hundreds of billions of dollars of oil that had flowed from the Niger Delta over the previous four decades. The oil industry, its Western client consumers and a strikingly wealthy minority of Nigerians seemed to do rather well out of a state that offered almost nothing to everybody else. It drove to distraction many of the Nigerians I came to know; one friend lamented that his nation had become a 'theatre of fraud'.

Once I moved to Nigeria and started reporting on it for my newspaper, I found myself drawn emotionally to this oil-ruined country's survivalist sense of mischief amid incipient breakdown. There is something universal in this, too, just as there is in the Delta's battles over crude. The country's modus vivendi seems to tap into a deep – and, in rich countries, mostly suppressed – human desire to subvert the rules and systems that prevent chaos but can also stifle creativity, ingenuity and spontaneity. Nigeria's turbulent history has forced its people to build a can-do culture par excellence.

For many foreigners and wealthy Nigerians, blessed with the twin luxuries of financial security and the fallback possibility of escape, the country's intensity becomes something to admire and crave. I felt self-conscious during my early days in Lagos, walking on streets where my white skin made me an oddity: children used to call out 'Oyinbo pepe!', a reference to a popular rhyme about foreigners who can't handle the spiciness of traditional foods such as pepper soup (which, it has to be said, can be nose-streamingly hot). But gradually I came to love the constant interest, the sociability and the generosity that runs like a skein of silver beneath the surface roughness of the place. I marvelled as people got by – and sometimes thrived – despite everything that seemed arranged to prevent them doing so. The flipside of the nation's abuses of power, everyday attrition and squalor was a compelling story of ceaseless, inventive activity. Like Tennyson's Ulysses, much had been taken from Nigeria, but much still remained. Its spirit was reflected by an ambivalent piece of graffiti that spoke like a ghostly social commentary from walls, websites and vehicles across the country: 'No condition,' it said, 'is permanent'.

The vital, worldly and intellectually absorbing Nigeria I gradually discovered seemed far removed from the remote, two-dimensional country that outsiders often imagined and dismissed. That Nigeria – in as much as it is thought about at all – is little more than a source of laughably unsophisticated internet frauds and periodic extreme violence, such as the rioting that drove the Miss World contest from the country in 2002. If there is a world-view of Nigeria, it seems mostly to mirror that of Colin Powell, the former US secretary of state, who once described it as a country of 'marvellous scammers', blessed with wealth it had 'pissed . . . away'. On the day I left London to move to Lagos, the woman at the British Airways check-in desk even told me she hoped I had my 'corrupt and criminal head on', in order that I might survive.

There are few countries as large as Nigeria that are so under-reported and little understood outside their shores. This ignorance carries a practical cost that has become clear to me over the years, as I have watched the tightening of the stifling bear-hug between Africa's giant oil province and the rest of the world. Nestled in its sweaty, pestilential geographical niche, in the corner of land formed by the junction of western and southern Africa, Nigeria is sometimes described pejoratively as the armpit of the continent. I prefer the topographical evocation once offered to me by an enraptured young man at an all-night church service in Lagos. It gives a sense of the importance of the place, its possibilities, and of its potentially deadly explosiveness. 'Africa is a pistol,' he told me. 'And Nigeria is the trigger point.'

That observation is as good a starting point as any for a journey into Nigeria's volatile relationship with the fuel that powers the world. It is a story of how the plundering of resources in the past echoes through the present day and rebounds – in the end – on those who have done the taking. This tale, which crosses continents and centuries, tracking the rise of oil as a global force, starts in the only place it can: the Niger Delta and its surrounding region, site of the encounters that first awoke me to the way the cult of crude gripping tracts of West African swamp reverberates through Britain and around the planet.

Back at the *agua petróleo*, Viera and Montero are as captivated as I am by the promise of liquid riches in the marsh around us. They are soon scooping the oil on to fallen leaves to pose for photos that I have trouble taking because my hands are so slippery. Once touched, the crude just seems to spread unpredictably and uncontrollably, coating my paper and clogging my pen as I try to write. There could be, Viera notes, *muito dinheiro* – a great deal of money – in all this mess.

Montero, skin glistening in the gloaming, says the money from oil can help his people to 'really move'. As he holds up hands caked with filth, he tells me that he's confident the wealth will be managed for the best. His reasoning both thrills and chills me, reminding me why the story of Nigeria matters so much, as a warning and a prophecy to us all.

'Nigeria,' Montero says, peering towards me in the fading light, 'has *experience* of these things.'

PART ONE

THE HUNDRED YEARS
OIL WAR

From Candles to Kalashnikovs:
an Alternative History of the World Crude Century

'If one finger brought oil, it soiled the others.'

– *Igbo proverb, cited by Chinua Achebe in*
Things Fall Apart

1

STARK ILLITERATES AND JUNKIES

As I looked across the water from Abonnema jetty, at the heart of the Niger Delta, I felt like a frontiersman standing at the edge of the known world. Behind me stood the tiny town, a waterside strip of buildings dominated by a neatly compact church and mobile-phone kiosk antenna that seemed to scratch like a giant finger at the sunless sky. Across the river from me were the creeks, where white egrets and other water birds patrolled the exposed mangrove roots and mud banks. I was at the barrier between urban development and the land of nature, remote villages and big oil beyond. From here, my journey – begun nervously by car two hours ago in Port Harcourt, Nigeria's oil capital – would continue by boat, jagging through waterways where oil pipes wended like submerged snakes.

I was tense because I'd come to meet Alhaji Mujahid Dokubo-Asari, self-styled scourge of the oil industry and emblem of the violent resistance that was increasingly coming to define the Delta in the eyes of the world. Earlier in 2004, he and his cadre of armed fighters, known as the Niger Delta People's Volunteer Force, said they'd begun an armed struggle for control of oil stolen by the Nigerian state and handed over illegitimately to the international oil companies. Protests against big oil in the Delta had gone on for many years, but Asari had stepped up the violence and the rhetoric. From his self-promotional talk, expertly fed into the international media, you would have thought he was Nigeria's Robin Hood, the creeks and mangroves his Sherwood Forest.

I'd already almost blown the chance to see Asari, whom I'd met for the first time a few months earlier. In a snatched mobile phone conversation the previous day, he'd told me to wait for him at the Abonnema jetty at eight in the morning. Unfortunately, I'd misheard him on the crackly line and gone instead to another jetty, called Abuloma, on the outskirts of Port Harcourt. He'd exploded when I'd called and told him where I was. 'What nonsense is this?' he had yelled, so loud that I had had to take the phone away from my ear.

As I stood waiting now, watching the wind whip up a swell, I wondered whether I had unwisely piqued a warlord. Then I noticed a speedboat making its way towards me. As it came closer, I could see it was filled with fit-looking young men. They seemed unworried about attracting the attention of the local people on the jetty who were shifting fish and other produce into boats. The new arrivals appeared a known quantity, their visit met with indifference.

Once the speedboat driver had moored, he beckoned me silently aboard. He lifted one of the footboards to reveal an AK-47 rifle hidden underneath. Then he opened his coat to show me a magazine of ammunition strapped to his torso. At all times he moved freely, without fuss and with little apparent concern about whether his weapons were seen by passers-by. When I asked if he was expecting trouble, he replied matter-of-factly, 'Any moment, at any time. But in this case, no problems.'

We cast off and headed for a wide-mouthed tributary off the river's opposite shore. Bouncing across the water, I felt a freedom and freshness far removed from Port Harcourt's dankness and constant traffic congestion. As we entered the tributary, I could see a further set of smaller waterways opening out, part of a network of creeks that branched like the stem of a bunch of grapes. Suddenly, the boat's commander pumped his arm with machine-gun rapidity to direct the vessel sharp right, down a narrow mangrove-flanked corridor. Everyone crouched to avoid overhanging branches, as the rapid movements and tight turns made the air around sing with a windy hiss. As we ducked and dived, I didn't know whether to be unnerved or reassured by the sticker on the hull that claimed our vessel was – like the *Titanic* – unsinkable.

After a few minutes, the branches fell away like a curtain to reveal the theatre of Asari's camp. I could see smoke and hear drums and chanting. A long white drape fluttered from a stick 20 feet high, like a pennant at a medieval English jousting tournament. The flag had been raised in honour of Egbesu, a spirit revered by members of Asari's Ijaw ethnic group. Nearby, a man in a red hat was holding what looked from a distance like a black chicken. I heard a sound like a flute, as if serenading the visitors to this sacred grove.

We came to a stop on a narrow strip of beach, where some men onshore helped pull our boat as far from the water as possible. Only then would they let me hop out. I noted that these militants, for whom shootouts with rival gangs were a fact of life, were surprisingly punctilious about making sure I didn't get my feet wet.

As I walked up from the tiny sandy cove, some of the dozens of young men who lived in the camp began to greet me warmly. I passed a machine-gun nest and a half-built outhouse, where breeze blocks for the next phase of construction were lying around ready to be used. Beyond was a long, single-storey building, like an army barracks, outside which several fighters were lazing around on mattresses. To my left, the ritual I had spotted on arrival was continuing behind some foliage, half-obscured from my view.

My host was watching the ceremony from his seat on the patio of the main building. He sat gazing thoughtfully, a Muslim chief watching his animist foot soldiers honour their religion. The picture of philosophical grandeur was qualified only by his appearance, which was more pimp than pasha: he was wearing a loose black tracksuit and T-shirt, and yellow flip-flops. His round face and generous belly sported the plumpness of privilege. He welcomed me with an observation that suggested I had indeed arrived in a new world. 'All here now is river,' he said, indicating the mangroves surrounding us. 'We are getting deeper and deeper into Ijaw country.'

Almost all Nigeria's crude oil comes from the swamps of the Niger Delta and the coast beyond, where the Ijaw live alongside the peoples of many other ethnic groups. The region is a landscape of the imagination that shifts from sandbars to mangroves to forest to field. Almost due south of the junction betweeen the Niger and the Benue, the two rivers that roughly define the country's northern, eastern and western regions, the Delta is an entrepôt of extraordinary biological and cultural diversity. Altogether, the watery fingers that branch off Nigeria's great waterways spread into a Delta region that is estimated to contain more than 10 million people. They speak many languages and are spread over an area that – by one commonly used measure – is about the size of Scotland.

During times of full production, Nigeria is the largest oil exporter in Africa and one of the top ten in the world. According to the US Government's Energy Information Administration, in 2006 it was one of half a dozen countries that had net average exports of between 2m and 2.6m barrels a day, compared with the mega-exporters of Saudi Arabia, which exported 8.53m barrels a day, and Russia, which exported 6.87m. Nigeria's peers at that time included Iran, Kuwait and Venezuela, all centres of great geopolitical interest.

The Delta's oilfields are explored and drilled by a group of foreign multinationals from the Western world and – increasingly – else-where. The largest operators are Shell, Exxon and Chevron of the USA, France's Total and Eni of Italy. Much of the crude is drilled and processed through unguarded wells and pipelines in the creeks, passing onwards through junctions known as flow stations to export terminals on the coast. The inherent vulnerability of these production and distribution networks is part of the reason the companies have begun to develop operations offshore, hoping that attackers won't fancy an assault on deep ocean fields of oil and natural gas.

Nigeria has a big strategic importance for the multinationals – and their home countries – because of the size and quality of its energy reserves, its geographic position and its government. Its crude is much prized because of the ease with which it can be refined into petrol, while its huge and still expanding proven gas reserves are already playing a big part in the worldwide shift towards the fuel. Nigeria is physically closer than the Middle East to the USA, reducing shipping costs. It is also more of a political friend. In 2007, Washington launched a special military command, known as Africom, partly to improve the security of oil reserves in Nigeria and elsewhere on the African continent.

Yet, for all the possibilities, Nigeria has lately become emblematic of the fragility of a world oil economy that depends on the movement of many millions of barrels of crude across oceans every day. Planned production increases in and around the Delta – which were supposed to take its output above 4m barrels a day by 2010 – have been hobbled in part by lack of government investment. Community protests and militants have also undermined the industry's existing activities, cutting production by almost a quarter by early 2008. Those kinds of figures are more than enough to alarm rich countries' governments and oil markets about the lawless 'state within a state' that the Niger Delta has become.

The modern, near-permanent state of Delta disruption, as exemplified by Asari, dates back to unrest during the early 1990s. Then, the people of a small region known as Ogoni had grown angry over pollution problems and other side-effects of oil. Protests prompted a shutdown of Shell's operations there, and a military crackdown. Scores of people died, according to human-rights activitists. The trouble continued and, in 1994, amid factional disputes within Ogoni, a writer and activist named Ken Saro-Wiwa and some of his colleagues were arrested for allegedly murdering four local chiefs. Despite a lack of credible evidence or judicial process, Saro-Wiwa and eight others were convicted and sentenced to be hanged by a special tribunal set up by the country's military dictatorship. The executions, carried out for maximum shock during a summit of Commonwealth leaders in New Zealand, led to Nigeria's suspension from that organization amid near-universal international condemnation of the killings.

The Saro-Wiwa executions marked the start of an era of deepening problems for the oil industry in the Delta, during which its legitimacy has come more and more under attack. Other larger ethnic groups – able to mobilize far more people than Saro-Wiwa – saw how effective the Ogoni protest had been in grabbing world attention. In 1998, Asari's Ijaw people put into words their anger with oil companies and the government. They gave the multinationals until 30 December of that year to pull out, 'pending the resolution of the issue of resource ownership and control in the Ijaw area of the Niger Delta'. 'We are tired of gas flaring, oil spillages, blowouts and being labelled

saboteurs and terrorists,' read their statement, adding, in a meaning-
ful nod towards Saro-Wiwa's fate, 'It is a case of preparing the noose
for our hanging.'

On the day of the Ijaw deadline, thousands of security force members were deployed to the Delta as protestors began to gather, some
of them bearing candles. Dozens of Ijaw people were reported killed
in the military crackdown that followed, with many more tortured
or detained arbitrarily. Since then, violence in the Delta has ebbed
and flowed, with the security forces carrying out notorious massacres
such as the murder of hundreds of people in 1999 in the town of
Odi, Bayelsa State, apparently in reprisal for the killing of a dozen
police officers. Olusegun Obasanjo, president between 1999 and
2007, showed no signs of outrage at the action and others like it. He
once told a journalist that another security force revenge massacre of
hundreds of people, after some troops had been captured and killed
in the central state of Benue, was an example of how 'cause and effect'
worked in life. 'In human nature,' he said, 'reaction is always more
than action.'

In such violent circumstances, it is hardly surprising that armed
militant leaders like Asari have flourished. In his meeting with me,
he proved a charismatic and fluent speaker, when he was not being
interrupted by the two mobile phones that lay on a crate in front of
his chair. Most of the time he talked calmly, occasionally wearing a
comically exaggerated frown that complemented his dry and cynical
sense of humour. When he was exasperated – as when he scolded me
for going to the wrong jetty – his voice rose in pitch and rasped with
the brittleness of dry leaves. As we chatted, I noticed his lips were
strangely mottled with white: the effect, he said, of an allergy to the
sulphur in drugs he was taking for malaria. The disease was one of
the hazards of a life in the bush spent avoiding the Nigerian army.
'Most of us don't sleep at night,' he said. 'We have to keep watch. The
soldiers always come at night.'

Living in hiding was part of a struggle that Asari claimed was
aimed at ending the suffering of the people in the Delta, particularly
the Ijaw, the region's largest ethnic group. The governance of the
region was 'abnormal' and 'fraudulent', he told me, spitting out the

derogatory adjectives like rounds from a rifle. The oil companies were 'evil collaborators' with the oppressive Nigerian authorities. Under a much-criticized land law passed in the late 1970s, all the revenues from crude production go to a central government account, with none passing directly to local people who live among the oilfields.

Asari said he wanted a sovereign national conference to be held to consider whether the Delta should be part of Nigeria or not. His logic would chill any oil executive or Western policy-maker who sees Nigeria's high-quality, plentiful crude as stable insurance against meltdown in the Middle East. 'For us, for me,' Asari said, searching for the right wording, 'the companies have no business being here.' It was 'for the people to decide' whether they wanted to work with the oil industry, he continued. The alternative for the multinationals was stark: 'They should go when a new nation emerges.'

By Asari's own account, he had made a strange journey to this covert life in the mangroves spent fighting the Nigerian state and the oil industry. He came from a privileged Delta family. His father was a high court judge; his grandfather, he claimed, a slave trader. He had a brother who was an academic in the USA, while one of his sisters worked for an oil company. In short, his personal connections seemed typical of the complex web of Delta relationships that link apparently antagonistic forces – like the multinationals, the government and militants – far more closely than outsiders might guess.

Asari said a key turning point in his life was when he dropped out of university and travelled the world. He found himself attracted by the revolutionary spirit of Islam, to which he converted in 1988. He spoke of a 'wonderful year' in Libya, where he said he met Charles Taylor, Liberia's warlord former president (later put on trial for war crimes). Asari said Osama bin Laden was one of his heroes, although he denied links with al-Qaeda and said he disagreed with bin Laden's methods. Asari saw his struggle as about oil, rather than as a proxy for a worldwide conflict between Islam and Western power.

Asari's religious ambivalence was clear when I asked him about the ceremony his men were performing. He said they were making a sacrifice to help protect them from dying in combat. As a Muslim, he didn't officially believe this worked, although he didn't dismiss the idea either. He said he had seen cases where bullets aimed at fighters had not entered their bodies. 'Even me,' he added. 'I have been shot before. I didn't die.'

Fighting involving Asari's militia and other gangs has plagued the Delta for years. The gangs wax and wane, carrying surreal names such as KKK, Icelanders and Germans, with new groups constantly springing from the ashes of the old. When I asked Asari why the militias chose names that were so strange and – given the racial implications of KKK – even nonsensical, he shrugged. 'Maybe because they admire imperialism,' he replied. 'It has to do with intellectual bankruptcy. I don't know. Most of them are stark illiterates and junkies.'

The other militias used drugs 'very well', he continued. They took cocaine, heroin, 'all sorts'. I asked him whether his men did the same. 'Maybe outside the camp,' he replied. 'But when they are inside, it's forbidden.'

I was sceptical of Asari's assertion that his gang was an ultra-disciplined fighting force compared with the rabble it was struggling against. He claimed his force was a mass movement with, at a conservative estimate, 2,000 people under arms. He had sent out 100,000 application forms to prospective members, he said – although he didn't have a sample paper to show me. The force had heavier weapons than I'd seen, including rocket-propelled grenades – although these were all 'at other camps'. The mountain of unverifiable claims reminded me of a remark a US journalist once made to me, in frustration at the difficulty of pinning down facts in a remote region of poor communications where rumours multiplied like tropical fungi. 'The sheer quantity and variety of untruths,' she told me in wonderment, 'are *incredible*.'

One particularly murky story concerns the origins of Asari's force. It was almost universally believed by Deltans that the regional government armed him to help rig elections in 2003 (this has always been officially denied). Asari was, unsurprisingly, vague on this point, although he didn't deny being a former supporter of Peter Odili, the state governor until 2007. If the common belief is true, then there is something grimly appropriate about the emergence of Asari's militia and others from a deeply corrupt political system. By striking out on their own, they have, like Frankenstein's monster, turned against their creators.

As we talked on, Asari showed no signs of becoming bored or of having anything else to do, even though he had told me he was expecting a Nigerian army attack. A young man brought him a tray of food and a big metal mug of Bournvita chocolate drink. He carefully mixed carrot, lettuce, corned beef and mayonnaise to make the filling for some sandwiches. He did not offer me anything to eat or drink. Between mouthfuls, Asari reflected on the essence of his struggle. What the world saw as Nigerian oil did not belong to the Nigerian state, he said. The oil was the people's and had been taken – illegitimately – into government control. 'They are still eating in Abuja,' the capital, he said, pausing between bites of his sandwich, 'building mansions with disregard to the people who live here.'

In the background, chanting had started up again. Then I heard a volley of gunfire. 'High spirits?' I suggested hopefully. Asari was apologetic, even a little embarrassed. 'Sorry,' he said. 'They want to test the potency of what they are doing.'

A couple of minutes later, a crowd of about 20 militants appeared, some of them carrying drums and rifles. As the beat started, they leapt around, waving their guns in the air in excitement. Even Asari started clapping along. A man dressed in an orange Shell jumpsuit, inseparable from his Kalashnikov, caught my eye. He and some others started wrestling, sprawling in the mud. Some of the photos I took turned out to be hilarious: the scene looked more like a reality game show with guns than the training base of a militia movement.

The increasingly riotous behaviour of Asari's men was making it harder and harder to concentrate on his political polemic. He explained to me that his force was funded through a mixture of subscriptions, donations from civilian sympathizers and money made from selling oil taken from pipelines. Just today, someone gave 3m naira (£13,600), he said, producing a small checked bag. He opened it to reveal six bricks of money containing a hundred 500 naira notes each, or a tenth of the total he claimed. He refused to reveal the identity of the donor, saying only that she was an Ijaw woman who lived in the USA.

As he showed off the money, a loud explosion cracked through the air. It was answered by a volley of gunfire. Then, after a short and nerve-racking pause, we heard some voices shouting, 'It's OK!' Again, Asari seemed unworried by what was apparently not an out-of-the-ordinary event. He said, 'They throw the dynamite to show the potency of what they are doing.'

By now, I had just about given up on extracting any more sense from a conversation that seemed destined to be perpetually interrupted. Instead, I sat back to watch the creek carnival that was unfolding in rain of steadily increasing strength. One of Asari's generals wandered by, ringing a hand bell. A well-muscled young fighter, wearing nothing but tight black underpants, started to move around jerkily, like a Covent Garden mime artist. Water ran to the ground off his bare chest. The whole atmosphere, charged with testosterone and a certain homoerotic tension, seemed more camp militancy than militants' camp. Whether I was watching spontaneous ecstasy or a performance for a foreign visitor was open to question, although I didn't get the sense of being much noticed until I started taking photos.

The men were gluttons for the camera. A fighter that Asari had identified earlier as a retired army major produced his rifle and posed, smoking a cigarette. Then he was swept off his feet and carried aloft by some of the other men. As he turned back to look at me, his eyes had a wild, drugged appearance. 'When he brought out his gun, everybody was afraid,' Asari explained. 'He can get *out of hand* sometimes. He's an old man.'

As the horseplay went on, Asari continually made and took calls on his mobile phones. He told me afterwards that he had spoken to a top policeman, a senior army officer and a rival militia leader, with whom he talked jovially in pidgin English about a possible peace deal. I had heard him conclude the call with what must be one of the more cheerfully delivered death threats in history. 'No problem,' he said. 'I will kill you. Thank you. Bye bye.'

While Asari fielded more calls, I went for a walk around the area. I soon realized this was less a highly equipped camp and more a jungle bolthole. Behind the main building, there was a desultory training area and a shrine to Egbesu, fronted by white flags and palm fronds. Down by the waterside, a little to the left of where we had arrived, was a jetty that served as a toilet. Turds were pocked everywhere on the mudflats, filling the air with a stench that made me retch as I pissed over the side.

When I returned to the main building, the men were still dancing, their enjoyment apparently unaffected – and perhaps enhanced – by the rain now sheeting down over the corrugated metal roof of Asari's hut. One young man cavorted in a paramilitary police helmet, with its kitsch official logo: a truncheon sprouting a pair of bird's wings. It was used more as a spoil of war than as protection, for the men said they believed it was Egbesu who really safeguarded them from death. One man, indicating the red cloth tied in a knot around his forehead, told me, 'With this, if they shoot me, nothing happens.'

As I again took my seat next to Asari, it was clear I had finally outstayed my welcome. He suggested taking the boat back to the mainland before nightfall. The time had come to leave him and his men to their world in which weapons, spiritual belief, ideology and mercantilism combined to such deadly effect. Asari offered me a final, credulity-stretching thought that he said showed the nobility of his cause. While only 17 of his fighters had been killed, his force had assassinated more than a thousand members of rival militias. 'How come?' I asked. 'We don't know,' he replied, thinking about it for a moment. 'We are on the right course, so we are protected.'

Before I left, Asari made me a tempting proposal. If I came back soon, he said, we could go to see one of the makeshift refineries where his group processed the oil it took from wells and pipelines. It was an exciting prospect, although I was sceptical about whether the trip would turn out to be quite as revealing as he suggested. No outsider had yet managed to penetrate the murky world of oil theft and – for all my journalistic self-regard – I doubted I would be the first.

As I travelled back to Port Harcourt, I had my first incoherent thoughts about a meeting whose significance would grow with hindsight. I could see Asari was someone to be watched, even if I couldn't yet predict the impact he and his rival ganglords were to have on Nigeria's oil industry for years to come. I was struck by how openly and confidently Asari called for his revolution of sorts. He clearly wasn't intimidated by the military; it confirmed the impression I had from my contacts with soldiers that the fear factor worked the other way around.

My visit had added to my dawning sense of how deeply rooted the Delta's troubles were, even compared with places I'd seen that were superficially more violent. Although I had never been a war correspondent, I'd been to countries in the grip of conflict, such as Liberia. Yet, frightening as these places sometimes were during the bad times, there always seemed the redemptive possibility of longer-term improvement. I rarely felt that in the Delta, which was alarming – paradoxically – because it never broke out into a purgative no holds-barred shooting war. Instead, it always seemed on the edge, wearing your nerves and filling you with foreboding as you listened to stories of grievances that had built up over decades.

My encounter with Asari had also begun to crystallize some of my emerging thoughts about the cost of Nigeria's oil. It would be disingenuous to say I was suddenly overwhelmed with guilt, but I certainly had an uncomfortable sense of seeing the veil lifted on something ugly that concerned me intimately. I could see the obscene asymmetry between the smoothness of my oil-fuelled life in Britain and the toxic impact of crude on one of its main source regions. Like other horrors that we tolerate in the West because they happen to people elsewhere, the disturbing story of Nigeria's oil became harder to ignore once it was no longer abstract. What had been faraway and theoretical had now become up close and personal.

Taking up Asari's suggestion of a return visit, I arranged to meet his men at Abonnema jetty the following week. There I waited again in the light of a muggy morning, as if trapped in some Delta *Groundhog Day*. A police car rolled by but didn't seem much interested in what two white people – I was there again with a colleague – were doing hanging out in a remote Delta town. Not for the first time, I had the sense that the authorities probably had a pretty keen idea of where Asari was, but didn't much fancy taking him on.

Asari's offer of a visit to his refinery was the best I'd had so far of an insight into a trade in stolen oil in the Niger Delta that is, by some accounts, colossal. By others it is simply very big. A confidential report commissioned by Shell concluded in late 2003 that as many as 685,000 barrels – or about a third of total production – were taken on average each day, generating between $1.5bn and $4bn annually for the thieves. Official estimates have run as high as 300,000 barrels a day, while oil industry figures have quoted lower – but still substantial – numbers of between 40,000 and 100,000 barrels a day.

The theft is a cosmopolitan affair, involving many parties to the Delta conflict. Asari claimed he sold the oil he took only to local villagers on a non-profit basis, although most outsiders thought that he in fact turned a tidy profit as a smuggling middleman. High-level officials are also suspected of being complicit in the business: two rear admirals were sacked in 2005 after a court martial found them guilty of involvement in the disappearance of an impounded tanker that had been shipping stolen oil. I once asked the commander of an army peacekeeping force in the region whether it was being carried out by local people, the armed forces, government officials or foreigners. He replied simply, 'All,' indicating that the meeting was at an end.

As before, Asari's men arrived to meet me by speedboat. I could see more weapons on board this time: a machine gun and three Kalashnikovs, lying behind benches on the floor. One of the youths apologized as he passed the machine gun over my head. The men were swigging gin, which blew in my face in a fine spray as we picked up speed. Again they said sorry – then one of them fired two shots in the air.

At the camp, Asari's force was in restless repose. One fighter worked out with dumbbells while others lolled on the mattresses, as if they hadn't moved since my last visit. A young man was reading, aloud in English, from a copy of *Macbeth*. He was studying a page near the end, where the murderous king is close to his own death. I read Macbeth's desperate words over the fighter's shoulder:

> *They have tied me to a stake; I cannot fly,*
> *But bear-like I must fight the course.*

Everyone, including Asari himself, seemed more skittish than last time. He disappeared at one point to change his clothes, returning wearing one of the orange Shell jumpsuits that seemed de rigueur in his gang. He put on a white hard hat belonging to Willbros Group, the oil services company. 'Do I look fine?' he asked me, running his hands over his ample stomach. Yes, I said, stifling a smile at the incongruity of it all.

Preparations for the trip were moving slowly, so I took a walk to the force's training area a short distance from the main camp. There, four men were smoking hash under a wooden shelter next to a patch of open ground. I knew from a rota posted on the walls of the main building that the fighters were supposed to spend six hours a day training, so I asked if there was anything going on that afternoon. One of the men replied that there was, but said they were resting for 'one hour, two hours'. The marijuana was to give them strength to work and fight, added Blessing Iblubor, another fighter. 'That's why we want to take it always,' he said.

I asked Iblubor what he had been doing before he joined Asari. He told me he used to buy petrol from filling stations to sell on the black market, a lucrative business during the domestic fuel shortages that – ridiculously – frequently gripped one of the world's top ten oil exporters. Like many other Nigerians, his life was spent trying to navigate – and profit from – a system that had caved in partly because of corruption. One main reason for the fuel shortages was that officials had allowed the domestic refineries to fall into disrepair, so they could claim a cut in lucrative fuel import contracts. 'I am not happy with government,' Iblubor said, stamping his foot. 'Because how can young men live in the bush like this?'

This longstanding contempt for government – shared by Asari, his men and huge numbers of ordinary Nigerians – had been entrenched in the Delta by the former president Olusegun Obasanjo's contentious re-election in 2003 to a second term in office. Many Deltans believe that rich nations, eager to project an image of progress in the country and to secure both oil supplies and the position of the Christian and generally pro-Western Obasanjo, all but ignored widespread ballot-rigging and intimidation of opposition supporters and their leaders, including the runner-up, Muhammadu Buhari, a Muslim northerner seen as less pliable to Western interests.

The elections in Rivers State, where Asari was based and where Obasanjo officially won almost 93 per cent of the vote, were among the worst anywhere. In a day spent travelling in and around Port Harcourt, I did not see a single person cast their vote legitimately. Instead, I saw ballot-box stuffing and intimidation of electors by ruling-party agents, and heard accounts of voting materials being stolen by armed thugs. In one counting centre, I watched as returning officers leafed through a sheaf of results sheets recording 100 per cent turnouts and 100 per cent votes for the president. In Port Harcourt, a group of young men identified by locals as ruling-party supporters tried to persuade me that a large street protest complaining about the non-distribution of ballot boxes was being staged by people who were mentally disturbed.

It all sat uncomfortably with the praise sprayed on Obasanjo by the rest of the world. Jack Straw, Britain's then foreign secretary, declared the elections 'a landmark in the advancement of Nigeria's democracy'. In private, a diplomat I knew gave me an alternative insight into what Western countries thought about the process. He had asked me, rhetorically, before polling day whether it would be 'too cynical' to suggest that he and his political masters 'might close our eyes and ignore flaws that are likely to be quite considerable in nature'. When I suggested that some people might indeed think this cynical, he caustically compared Nigeria's condition to that of another Western-allied oil state with a notoriously poor record on democracy and human rights. 'When we insist on the Saudis having free and fair elections,' he said, 'come back and complain about the Nigerian imperfections.'

By late afternoon, Asari was finally ready to leave the camp for the refinery. The weather was bad and we spent the first part of the hour-long speedboat ride crouched with heads down, trying to stay out of the rain. Asari had said we would probably see a military base and oil thieves' barges out on the river, but we came across only a few fishermen in canoes. Beyond a stretch of blackened mangroves – burnt by a recent fire at the refinery, Asari said – we arrived at a gap in the vegetation across which some fallen trees lay. A thin green pipe was just visible above the waterline. Asari said that this was where his men tapped the oil, but that unfortunately we could not go further to see the refinery, because the tide was high and the route blocked. It seemed surprising that an Ijaw man apparently so conscious of his roots and environment should be caught out by the predictable daily movements of the Delta's waters.

On the way back, we passed Shell operations where two Dante-esque orange plumes of burning waste gas illuminated the overcast sky. Another company facility, with distinctive yellow and black railings, stood apparently deserted, a ladder leading conveniently down to the water. There appeared to be nothing and no one to stop us climbing up ourselves, emulating the gangs and community protesters who periodically occupied facilities and disrupted production all over the Delta.

We made a final stop at the nearby community and Asari stronghold of Sangama. We left our soaked-through footwear at the doors of the neat, tiled front room of the community chief's house, where some of the chairs were still encased in their plastic covers. A large television with the volume turned up high was showing an action film starring Sylvester Stallone. Boma Briggs, a community leader, explained that oil production had brought nothing except pollution, making it harder to catch fish. The area lacked clean drinking water, network electricity and school and medical facilities, he added.

Briggs talked about the problems of conflict and security in the area, explaining that local fishermen voluntarily paid protection money to community leaders. The cash was put into an account administered by a local commander, who gave donations to those in need. At the back of the room Briggs showed us evidence of some of the proceeds from the arrangement: four Panasonic television boxes and one for a JVC sound system. 'All these are gifts,' he explained. 'To say, "Thank you for your good security work."'

As we left, dusk was approaching and members of a crowd of dozens were chanting, smoking and drinking in boisterous appreciation of Asari. Clutching his blue Beretta pistol, Asari sang a motivational Ijaw song to the gathered men, who replied in kind.

Then we piled into the boats and sped back to land in disorderly convoy, the drivers demonstrating their skills with sharp, fast turns that more than once threatened to capsize us. One or two of the fighters whooped, enjoying their power and freedom on the deserted waterway. One boat flew a Jolly Roger, while another was guarded by a fighter in a ripped leather jacket, black beret and machine gun.

Before Asari and I parted, I asked him what would happen as his campaign developed. His reply was thoughtful and troubling. 'One small mistake can bring everything down,' he said. 'Like all revolutionary movements, such mistakes can be very expensive. It can lead to the deaths of so many people. So we are trying to prevent that.'

Within months of uttering those words, Asari's words were being put to the test, as he captured the world's attention in a spectacular coup de théâtre that must have exceeded even his greatest expectations. He threatened to launch an offensive called 'Operation Locust Feast', unless government troops backed off in the Delta and the authorities began talks about oil resource control. He denounced the oil multinationals and said he could not be responsible for the safety of foreigners working in the area. His timing was spectacularly good, helping push the world price of crude through the psychologically significant barrier of $50 a barrel for the first time. Shell evacuated almost 250 staff from Delta locations. Asari flew to Abuja, the capital, to broker a peace deal with President Obasanjo under which he agreed to disarm in exchange for vague commitments to giving the Ijaw greater self-determination and control of the oil wealth on their land. Obasanjo condemned 'undue militancy' in the Delta, although he admitted the region's people had legitimate grievances and he criticized local officials for failing to bring development. 'The obvious assessment so far,' he said, risking no overstatement, 'is that not much impact has been made on the lives and living standards of most ordinary people of the Niger Delta.'

Operation Locust Feast and the deal that followed it made Asari an international icon of the new generation of gangster-rebels holding sway over large parts of the Niger Delta. His force might have been disorganized and have exaggerated its own importance, but it had the guns to make its point. Asari himself epitomized the contrast between the militia's operational chaos and its political muscle. At times he seemed almost a vaudeville figure, aware – at least to some degree – of his own inconsistencies. Yet he was clearly a ruthless leader of men and skilled user of activists and journalists to help publicize his cause. In a system so riddled with corruption and contradictions, perhaps it was no surprise that someone like him could appear to be both populist freedom fighter and profiteering warlord, feeding off the scraps of the very oil industry he was trying to undermine.

I managed to see Asari one more time, several months later. My appointment with him was in a courtyard outside a law firm's office in Port Harcourt, where some of his former militia members were also waiting to see him. Much had changed since we had last met: the peace deal meant Asari and his fighters had come out of the bush and were supposed to be reintegrating into urban life. Now many of the men were restless and bored, complaining they had no work. One slept on a car bonnet, while others asked me for money for food.

As dusk approached, Asari swept in, wearing white robes that had a preternatural appearance in the fading light. Addressing the young men from under the mess of power and telephone lines that hung precariously over us in this urban jungle, he called out a list of names of people to come forward and stand in a line. In the background, one of his assistants was distributing gifts, including at least one watch supposedly made by Calvin Klein in Switzerland (although probably plucked from the constellation of counterfeit designer goods sold at markets across Nigeria). The men picked would work for Asari's new security company, which he later told me had contracts with oil industry businesses he wouldn't name. One of the chosen militants looked ecstatic: he made a thumbs-up sign to someone in the crowd and gestured to his eyes as if to say, 'Can you believe what you are seeing?'

Afterwards, Asari handed out cash to some of those not selected. 'Money, money, money,' one of them said aloud in his excitement. It seemed an *Animal Farm* moment, sealing the transformation of Asari from renegade to potentate.

As it was now dark, Asari suggested we talk in the back of his Lincoln Navigator luxury four-wheel-drive car. When I asked if the vehicle had a television, Asari pulled down a screen that was showing a poor-quality broadcast of a female presenter talking. The car needed some maintenance, he said, because it had been parked up for a long time while he was in the swamps. He told me later that he owned seven vehicles: the Lincoln, four Mercedes and two buses.

The strange circumstances of our meeting were heightened by the presence of a secret police officer detailed by the government to stay with Asari. The officer sat silently on the back seat and did not interfere; the few comments he did make suggested he sympathized more with Asari than with his government paymasters. It wouldn't be the last time in the Delta that I was struck by how some of the most feared public faces of the state turned out to be among its harshest critics in private.

Asari, too, was unhappy with the authorities, despite his apparent détente with them. Many of his bases, including the one where I'd visited him, had been damaged or destroyed by the security forces, he said. He was angry at the official failure to address the political dimension of his struggle, including demands for greater control of oil wealth and self-determination for the Ijaw. These questions were touched on in the peace deal, although no specific commitments had been made to address them. I left with Asari's menu of threatened retaliatory action – 'civil disobedience, [oil production] stoppages, demonstrations, boycotts and so on' – ringing in my ears.

Asari's demands for the authorities to address his grievances were curtailed abruptly. Months after I saw him for the last time, he was arrested for treason. I followed the twists of the case in the newspapers, always with the thought that it was unlikely to be the end of the story for this most maverick of militants. Asari, just like the Eli Wallach character in *The Good, the Bad and the Ugly*, seemed the kind of man who would always escape just before the noose pulled tight.

During Asari's unplanned absence, the next episode of the Delta's increasingly deadly drama began. A new group called the Movement for the Emancipation of the Niger Delta (MEND) started a series of car-bombings, attacks on oil installations and kidnapping of oil workers. By late 2007, when Asari was released without ever going to trial, MEND had turned into one of the most acute of the militant threats to the oil industry worldwide. Asari plunged straight back into the Delta's conflict politics, adding another ingredient to an already dangerous mix.

Watching the rise, fall and re-emergence of Asari, it struck me that he, MEND and the other warlords were far from inexplicable mavericks who had sprung out of the mangroves as if from dragons' teeth. They were, on the contrary, an inevitable product of an industry that had over decades come to engulf a country. They were part of a rolling conflict over oil that had begun not with Operation Locust Feast, or even with Ken Saro-Wiwa's hanging, but more than a century before. As the troubles had endured and grown, more and more Nigerians had – as Asari himself put it to me – decided it was time 'to assert themselves and take what belongs to them'. As I travelled deeper into the Delta, I was starting to understand what that meant in practice – and why it concerned many of us outside the country far more than we might ever have imagined.

2

WHERE DUTY AND GLORY LEAD

Drive west long enough from Port Harcourt, along roads flanked by long curtains of elephant grass, and you will cross into a small but storied state named Bayelsa. Defined by a coastline that curves like a lithe Delta barracuda, it was created in the mid-1990s mainly as a homeland for the Ijaw people who fish the region's waters. Perhaps because of its modest size and youth, Bayelsa tries a tad too hard to assert itself. Official number plates in the country's equal youngest territory proclaim that it is already the 'Pride of the Nation' or – still more ambitiously – the 'Glory of All Lands'.

One of the reasons for Bayelsa's apparent self-confidence is that it can claim to be the birthplace of Nigerian oil. It was here, in the felicitously named town of Oloibiri, that Royal Dutch Shell made its first commercial find, in 1956. Two years later, senior Shell and BP officials waved farewell to the first export cargo of crude, destined for a refinery on London's River Thames. It was the genesis of an industry that would play a huge role in defining a country, its people's fortunes and its place in the world.

As I approached Oloibiri one typical Delta morning already thick with sticky heat, I passed an anonymous-looking clearing hacked from the jungle. A rusty barbed wire fence surrounded an equally dilapidated oil-wellhead, known in the industry as a Christmas tree because of its branching network of pipes and valves. The accompanying signboard, rendered barely legible by corrosion, was as understated a historical landmark as you could wish to find anywhere: 'Oloibiri: well number one,' it read. 'Drilled June, 1956. Depth, 12,008 feet.'

A few yards away, behind a pile of wooden planks, a black-tiled monument commemorated the March 2001 visit of President Olusegun Obasanjo. The dedication announced the laying of the foundation stone of the Oloibiri Oil and Gas Research Institute, 'to the glory of God and service to the Niger Delta people of Nigeria'. The institute, like so many projects in this region of unfinished business, had been promised but not built. A local official who was with Obasanjo on the day of the dedication told me the president was much irritated when he found out he was to inaugurate a project for which no funds had been made available.

As I stood and watched, a local man pulled up alongside me on a motorbike. He introduced himself as Giedia Dangosu, an Oloibiri resident. When he saw what I was looking at, he told me a little of his family history. His father had acted as a mediator between Shell and the local community, he said, as well as working as a tea-boy and clerk for a British colonial district officer. Asked how he felt after five decades of oil production in Nigeria, his reply was unequivocal and immediate. 'Cheated and brainwashed,' he said. 'I don't know the proper language.'

Dangosu invited me to follow him as he continued his journey down a paved road that stopped suddenly a short distance further on, outside Oloibiri town. There, it became a dirt track into which vehicles sank during the Delta's frequent prodigious rains. Another man came to greet us, introducing himself, in one of the many small signs of big oil's corporate cultural incursions into the region, as the town's public relations officer. When I asked how he was, his initially ebullient reply slipped gradually into a controlled, almost wry, despair. 'Fine, fine,' he said. After a short pause, he continued, 'In fact, I can't say fine and sit here and be dying. We are dying. Not fine, in fact. But we are managing.'

He asked me to wait a few moments for an audience with Chief Osobere Inengite, the town's traditional ruler. I was shown to a wooden bench on a concrete patio that gave onto one of the many dirt tracks crisscrossing the village. Behind me was a small house whose gloomy hallway was impenetrable to the eye beyond a few feet. This interior darkness is common in Deltan houses, where corrugated metal roofs extend like carapaces to protect unglazed windows from downpours.

The chief arrived, dressed smartly in a traditional black Deltan suit with a metallic silver sash that hung diagonally across the chest like a military honour. He was a dignified but slight man who seemed as if he might crumble to dust at too careless a touch. He told me he was 72 years old, although his slow movement, measured speech and parchment skin made him seem more ageless oracle than village elder. He listened carefully as I told him I had come to learn a little more about the origins of the oil that now shapes Nigeria's destiny.

When he replied, he spoke with the manner of someone who had thought and talked often about a subject at the forefront of many Deltans' minds. Inengite's story was of the great arc of the oil industry's involvement in his community, from exploratory promise to shutdown and a sense of betrayal. He said he wanted Shell's share-holders and chairman to come to his town to see what the progenitor of Nigeria's wealth had become. Oloibiri should have been like Texas, or Johannesburg after the discovery of gold, he observed bitterly. Instead, big oil had taken what it wanted and in return polluted rivers, devastated land and offered little compensation. 'They don't only steal from us,' he said. 'They are also out to kill us.'

The chief dispatched someone to fetch a copy of what he said was the first contract the town agreed with Shell. Signed in January 1956 by two Shell managers and two community elders, the papers gave the company the right to the use of 2.138 acres of land for five years at a rental fee of one pound per acre per year. The document had originally referred to the payment as a 'valuable consideration', but this had been crossed out by hand and replaced simply by the word 'compensation'. It didn't seem much to hand over for what was to become a prime bit of real estate, where oil was pumped until the 1970s. When asked, Shell said it had no record of the Oloibiri contract, adding that it had lost many documents during Nigeria's 1967–70 civil war.

On the threshold of the 50th anniversary of Nigeria's first commercial oil exports, the people of Oloibiri saw no reason for celebration. Inengite said crude had despoiled their delicate fluvial habitat. At first, the villagers had creamed their skins in oil slicks, not realizing that the strange and fascinating liquid was a toxic substance. Then fish began to die, along with plants that produced staple foods such as cocoyams.

Oil also brought tensions with other nearby communities, one of which still contests Oloibiri's claim to own the land occupied by well number one. Chief Inengite said his town's control of the area dated back to the nineteenth century, when it conquered the 'hostile people' living there. As in many other resource-rich regions throughout the poor world, this is a dispute without written records and therefore without obvious end. The chief shrugged at his opponents' obstinacy. 'Because we now have no big men, some people have come to distort history.'

As he spoke, there was a rumbling of the thunder long threatened by the gunmetal sky. I noticed that we had attracted a small crowd as we were chatting. All of us were male, brooding impotently, from our various perspectives, on affairs beyond our control. Across the path from us, just a few yards away, some of the village women were cooking and cleaning, making sure life went on while the men put the world to rights.

Chief Inengite expressed anger that Shell had helped the leaders of other communities in the area, but not his own. The company should be providing lifts in its helicopters for him and the other village elders, he said. It reminded me of a conversation I'd once had with a traditional chief from another Delta town, who, after describing persuasively the damage oil had done to his community, grumbled that Shell had not even given him a mobile phone for Christmas. I often found it hard to tell with Deltan local leaders where their ambitions for their communities ended and where their personal desires began.

Once the long-promised rain started to fall, I huddled with Inengite on the covered part of the patio nearest the house. The view was spare, as we sat on the bare concrete, looking out at the wall of water we could hear peppering the metal cowl above us. The chief, his rhetoric darkening with the sky, told me that expatriates like me in the Delta should know the terrible things 'a Nigerian can do to his fellow man' when under stress. Prolonged disappointment and post-colonial generational change had bred Delta youths who were angrier, louder and more assertively militant than their fathers. 'None of us, old and young, big and small, is happy about this,' he said. 'But we are happy that, unlike yesteryear, we have people who will shout, shout and shout for us.'

It is often hard to convey to outsiders the degree to which the story of oil permeates the psychological warp and weft of Nigerian everyday life. The pronouncements of officials from Shell, Chevron and the other multinationals are often front page news. Big oil's sometimes bewildering jargon – such as 'illegal bunkering', or taking oil offshore without official permission – is almost part of the vernacular. It's a rational national obsession, born of statistics and economic dominance: during my time living in Nigeria, oil sales typically accounted for about three-quarters of government revenues and more than 95 per cent of export earnings. In Britain, where production levels are similar, crude accounted until 2006 for less than one-tenth of exports.

In the Delta, oil is even more of an overbearing force. It is there in the waterside signboards that remind fishermen and other creek users to watch out for underwater pipelines. It is there at the export terminals like Bonny Island, where the multinationals' squat cylindrical storage facilities sit alongside each other like outsize cans of tuna. Most strikingly, it is there in the oil slicks stifling the delicate mangrove habitat, and in the towering orange flares of waste gas that bathe large parts of this electricity-starved region in a ghastly night-time half-light.

The oil spills the Delta suffers make California's La Brea tar pits – which once swallowed woolly mammoths whole – look like innocuous puddles. In January 2008, Nigeria's National Oil Spill Detection and Response Agency said it had found more than 1,150 oil-spill sites abandoned by various oil companies in the Niger Delta. Between 2003 and 2007, Shell alone admitted it had suffered well over 1,000 oil spills, adding that the majority of these were due to sabotage. Once again, the routine awfulness of Delta oil pollution is hard to appreciate unless you've actually seen it. When I showed some pictures of a slick at least the size of a football pitch to a well-informed and worldly friend in London, he asked ingenuously: 'Does Shell know about this?'

The pollution is one of many grotesque – and often absurd – manifestations of big oil that would not be tolerated in the countries where the multinationals are based. On one trip to Shell's Eriemu oilfield, I was perplexed to see a group of local women clustered uncomfortably close to a roaring gas flare that towered many feet above them. When I came closer, I saw they had arranged scores of wooden racks in a circle around the fire to dry their crop of tapioca, a white grain derived from cassava. The area, which was easily reach-able by an ungated footpath, had a warning sign, presumably put up by Shell: 'Danger!!! Entering of the flare area by unauthorized persons and tapioca drying at your own risk.' The women, perspiring even under the umbrellas and handsome wide-brimmed sunhats they had bought to protect themselves against the heat, were pragmatic about their working conditions. 'We are aware that there may be some health implications,' one told me. 'But we don't seem to have any clinic.'

The woman's unhappy aside is part of a lament heard in Oloibiri and a galaxy of other Delta communities transformed by oil produc-tion over the past half-century. If it had been in the Western world, Oloibiri might have been a city built on the wealth of the oil it sat on, a Houston, or at least an Aberdeen. Instead the motto of its gram-mar school – 'sacrifice, sweat and success' – seemed like a cruel joke in a community where getting a fair deal from life was far more elusive than that mantra suggested.

Like many other Delta towns and villages, Oloibiri was on the surface a place of picturesque tranquillity. Its network of paths was studded with periwinkle shells, the building material of choice across this estuarine region. The thoroughfares led through tall grass and past lines of mango trees, their leaves divided into green peninsulas that were round and fat like children's fingers. The church was neatly kept, its orange pastel walls offset by a brick red roof. The scene was bucolically seductive for a visitor well fed and healthy from the urban comforts to which he knew he would return.

Chief Inengite dispatched his nephew, Sunday Nyingife, to show me some of the darkness beneath the pastoral agreeability. A fit, stocky young man of 30, dressed in a close-cut white T-shirt, Nyingife could have walked off the pages of a Ralph Lauren catalogue. He was a sartorial and physical counterpoint to his spirited but fading uncle.

Nyingife stopped first outside a large, padlocked building set in a patch of grass. Its cream paint was peeling away to reveal grey walls and a rotting roof. A plaque on the wall said the structure was a six-classroom block donated by Shell in 1992 on behalf of the company and its joint venture partners, the Nigerian government, Total of France and Italy's Eni. 'They just built it,' shrugs Nyingife, 'but up to now, there is nothing going on. Since then, no lessons, ever, no teachers employed.'

Heading to the waterside, we passed a toilet block that consisted of a series of holes directly over the river, like the latrines that used to open directly onto the streets of medieval England. Two empty tins of Titus-brand sardines had been trodden into the path. Nyingife turned on a pair of taps that stood nearby. Nothing came out. 'No water,' he said. 'Abandoned projects.'

Oloibiri's deprivation was more subtle than the raw poverty of a remote village where people live in hovels and practice subsistence farming. As in other Delta communities that have relationships with the oil industry, it was clear that some people in the town were making money. Nyingife pointed to a large house owned by a prominent local who was away visiting South Africa. Crude had brought riches to a few, a wealth amid want that made the lack of public services for the masses seem even more obscene.

At the town's edge, Nyingife dived into a house and led me down a hallway. I passed by several silently inquisitive girls. Off the corridor was yet another darkened room, where an old man was sitting. He was introduced to me as Clifford Inengite, Nyingife's grandfather, who had provided me with the copy of the 1956 exploration contract with Shell. Inengite said the two parties to the deal communicated solely through a translator, the only man in the area who knew both the local Izon language and English.

After a short chat, we discussed whether Inengite would like to be compensated for providing a copy of the Shell agreement. The old man replied with dignified subtlety, implying obligation without saying so explicitly. He could not demand anything, he said: any gift must be freely donated. It was clear what had to be done.

As Inengite took his cash, my guide Nyingife broke unselfconsciously into a chorus of 'Money, money, money'. My encounter with Asari's militia man flashed through my mind. Nyingife's boyish giggling, a coarse contrast to his grandfather's stateliness, seemed to say something about how attitudes in the Delta had changed after five decades of oil production. Unlike the older man, Nyingife and his generation had grown up tantalized by the rich man's world.

Outside, Nyingife added a final, ambivalent postscript to the tale of exploitation he had recounted as we walked round the village. Standing at the boundary between the village track and the paved road, he revealed without prompting that he had helped to rig the last two elections on behalf of the ruling People's Democratic Party. It was the only route to power, he said: all other parties were a 'waste of time'.

'If you can't beat them, join them,' he said cheerfully. 'That is what is going on.'

It was a depressing valedictory thought to hear in the spiritual home of oil. It was also entirely typical of what many young people were saying in a region whose fundamental problem was not precisely mass poverty, but the marginal richness that oil brings. As long as youths could see a few local politicians – or militia leaders – dressing well, living comfortably and driving fast cars, the understandable temptation was to try to emulate them, or at least to hang around waiting for a few leftovers from their tables. A well-off Nigerian friend once described to me how he was struck by what he interpreted as a look of envious admiration shot at him by a harassed bus passenger gazing down at his air-conditioned car. 'He didn't want to kill me,' my friend observed. 'He wanted to *be* me.'

The more of these stories I heard, the more I realized I was holding a magnifying glass not just to the Delta, but to the sensibility of my own society and that of the other consumer states of Nigerian oil. Already I could see many common themes. The rich men's houses in impoverished Oloibiri were no more obscene than their counterparts in London; their opulence was simply starker compared with the general standard of living. Nor were the Delta resource control disputes so different in essence from the campaign of the Scottish nationalists for dominion of the UK oil pumped off Scotland's shores. That, too, can be cast as a story of historical oppression and growing resistance to a nation state that some see as an unwanted fiction. In Nigeria, guns and deeper poverty have simply bolstered the polemic.

There was a puzzle amid the parallels, too: why had Oloibiri taken such a material low road compared with the country that had received its first oil in 1958? A year earlier, Prime Minister Harold Macmillan had claimed the British people had 'never had it so good'. In late 1950s Britain, the austerity of the postwar years had – for some – given way to a period of steadily growing prosperity, built in part on the growing world oil industry. My parents – then ten years old – were part of a generation in which many British people would enjoy unprecedented improvements in their standards of living. I would be able to study at university, supported by a family that had previously had little contact with higher education.

Nigeria and Britain had never been on a common development path, but the divergence between the two countries during the Delta's oil era was stark. It seemed glib to blame Nigeria's troubles on the bad influence of crude, and also patronizing in its dismissal of people's ability to shape their own futures. Yet the oil industry was clearly such a determining feature of the nation's life and inter-actions with the wider world that its still-evolving role needed to be explained and understood. That meant going back even further in time and travelling across Bayelsa State to find another of the many histories of oil hidden in this littoral enclave.

On the river to Akassa, a town perched like a periwinkle on the Delta coast's southern apex, life and death seemed to emerge with the sun's first light. In the shoreline shadows cast by raffia palms, I saw flashes of colour as fishermen flitted in and out of view, as elusive as the butterflies of the surrounding forest. A woman in white bent double as she sorted produce into a wicker basket, neatly mirroring the form of an egret pecking morsels from the ground nearby. A bloated animal corpse floated by, its identity lost to decay: it might have been a small deer, or a large bush rat.

As my boat moved into the open waters, the atmosphere sharpened and the lingering sense of sleepiness disappeared. The wind blew harder and the estuarine swell ahead looked dark and threatening. This place was a watery amphitheatre, a great space that freed the mind but which had been used as a means to abuse bodies. Beyond lay the pounding eastern Atlantic waves that had claimed the lives of so many of the slaves taken from West Africa by the British and other Europeans.

I was nearing my goal: the site of perhaps Nigeria's first war over oil. In 1895, local people near Akassa had rebelled over the way the British ran the buying and selling of palm oil, a lubricant extracted from distinctive arachnid trees found all over the region. The traders, muscled out by Britain's Royal Niger Company, had risen up, just as their twenty-first-century successors would take to the swamps in armed protest against oil companies and the Nigerian government. The nineteenth-century revolt, by the people of nearby Brass, had triggered a reprisal attack that devastated the town and killed somewhere between two dozen and two thousand people.

The crushing British response was an early indicator of both the importance to the industrialized world of Nigeria's oil supplies and the lengths to which it would go to protect them. Viewed from the present day, the Royal Niger Company's experiences seem to foreshadow the state of the Delta region more than a century later. The conglomerate's 13-year hegemony was characterized by conflicts over resources, security force brutality and an influx of weapons, culminating in a violent uprising against the central authority. It was the first part of an ongoing Hundred Years War over oil that – like its counterpart more than five centuries before – has flitted in and out of peace before reverting inexorably to conflict.

Landing in Akassa, I was met by a group of men including Richard Grimu, appointed by the local community to be my guide for the day. An intelligent, reserved man, whose gentle manner and apparent fatalism masked a streak of righteous anger, Grimu led me down the shoreline, past mangroves hopping with mudskipper fish to a place known in the local Izon language as 'White Land'. An old warehouse stood to the side of a path, along which local people biked between villages; beside the thoroughfare, villagers laid out basins of prawns to dry in the sun.

Symbols of colonialism were scattered around, including a couple of cannons, an old railway track and metal rings embedded in the quay for mooring ships. A corroded sign propped against a shed bore the distinctive 'Y'-shaped emblem of the United Africa Company, the Royal Niger Company's forerunner. Each of the arms carried with it a Latin word: *ars*, *jus* and *pax*, or, roughly, arts, justice and peace. The company, which used to be part-owned by Britain's Unilever, one of the world's biggest consumer businesses, records the true meanings of these superficially inoffensive mottoes with surprising candour on its website. *Pax*, it says, stood for 'peace and order', which the company evolved to stem the 'anarchy and barbarism of the Niger Territories'.

Around the corner, at the Royal Niger Company's old headquarters, the corrugated metal walls of the ground floor were corroded beyond repair and the upper floor had disintegrated. A once sturdy safe in the corner was a mess of stone and mangled metal. The building's only occupant was a bare-breasted old woman, made pitifully thin by age and long deprivation. She sat eating from an orange plastic bowl and begging visitors for food and money.

Grimu told me that, after independence, these headquarters became a secondary school, then a shop, before falling derelict. The dilapidation was a sign, he said, of the decline of a community that had been designated special by the British because of its importance in the palm oil trade, but was now just one anonymous, impoverished village among many in the Delta and elsewhere in Nigeria. Akassa, 'first among equals' in colonial times, according to Grimu, was one of the imperial-era fossils strewn across the world, exploited, discarded and ignored. 'Why should Akassa now not be in the books of Nigeria?' Grimu asked, more baleful than furious. 'Why should it not be on the map of Nigeria?'

A little further inland, Grimu showed me another revealing relic of British occupation. In the forest, set back from the United Africa Company's facilities, was a clearing, fringed by oil palms and filled with bird noise. The boundary was marked by a line of thin-trunked and narrow-leaved trees. They formed a kind of natural palisade, funnelling the morning sun into a whirlpool of light that played on the centre of the glade. Round the sides of the plot, untended and un-illuminated, lay British graves, their white marble plinths blackened with fungi and their wooden crosses rotted. Most of the memorial inscriptions, dating mainly from the 1890s and 1900s, had worn away. One that had not was dedicated to 'our dear brother' Henry Russell, who died on 1 March 1895, in the midst of the conflict between Britain and the people of Brass.

Suddenly, one of Grimu's colleagues shouted out 'Jesus!' The revelation was material, rather than spiritual: he'd just knocked the top off one of the fractured crosses, which he was now hurriedly try-ing to rebuild. It had broken the crepuscular spell. The birds' chatter now seemed a mocking commentary on the Ozymandian aspirations of colonialists who thought they would rule in perpetuity.

If the graveyard's narrative seemed a familiar one of imperial intru-sion and oppression, Grimu's interpretation of it came as a surprise. Instead of viewing the British as rapacious aggressors, he talked of them as incomers to Akassa who were tolerated and even welcomed. His remarks were an unsettling illustration of how imperial policies of divide and rule cascade down the years on the tongues of local storytellers, setting neighbouring communities against each other as they compete to make their differing versions of history the universally acknowledged truth. Grimu was withering about the revolt against the British by the people of Brass, Akassa's neighbours across the water. He said, 'They came and raided our guests, who were trading on our soil.'

The residents of Akassa did not view the British who ruled here as cruel, Grimu continued. Instead, they saw them as visitors who 'beamed light' on the community. Unfortunately, the present government was keeping the village in darkness, he complained. 'The Akassa man has been marginalized and neglected.'

Grimu stopped suddenly to give way to a woman who was listening in and wanted to speak. She didn't want to let his account stand unchallenged. For her, his comments had grown just too parochial and self-pitying. In a country full of claustrophobic local feuds, she offered a broader perspective on a history moulded by the Royal Niger Company and other British interests. 'Every Nigerian,' she said softly, 'has been marginalized.'

My tour with Grimu had its genesis in some revealing research on the Delta I'd done in Britain. On a series of long, hot summer days, I'd hunkered down over boxes and bundles of colonial-era memos at the National Archives near Kew Gardens in London. This well-groomed western suburb of the capital was an uncomfortably benign and civilized backdrop to the events described in the papers I was reading. Over lunchtime sandwiches in the Archives' gardens, watching the birds clustered on the artificial lake, I reflected on what the documents revealed about the origins of what seemed a perpetual international struggle for Nigerian oil. So many of the reports seemed to point to the reverbatory significance of the colonial period and of one event in particular: the 1895 Akassa conflict and its aftermath.

The Royal Niger Company's dominance of Akassa and other tracts of the Delta was the culmination of several centuries of European plunder in the region of West Africa that became Nigeria. The first contact came via the arrival of the Portuguese in the late fifteenth century. Soon, trade in gold, other commodities and slaves flourished. The Portuguese, British and Dutch fought over territory they could already see was rich in resources and ripe for exploitation. Local chiefs proved willing suppliers for the Europeans' hunger for human beings.

Then, in 1807, as a result of campaigning by abolitionists and religious sentiment, as well as changing commercial times, Britain outlawed slave trading by its own nationals. As part of its plan to end the practice altogether – thereby, among other things, preventing its European rivals gaining a competitive advantage – London deployed patrols on the West African coast to intercept slave-trading vessels, free their cargoes and arrest their crews. In 1851, British ships attacked Lagos, then a small island-port, on the grounds that it was being used for slaving. A decade later, London formally declared the area a colony, gaining for the first time a political foothold in the landmass that was to become Nigeria.

As Britain grappled for political control, its companies fought for commercial victory. Their main rivals were the French, who were eventually to take charge of many countries across the West Africa region, from Senegal to Cameroon. All the businesses wanted power over the River Niger, an invaluable trading route that rises in Guinea and snakes through the region before emptying into the sea off the southern coast of the Niger Delta.

In 1886, the United Africa Company – by now a behemothic agglomeration of British manufacturing and trading interests – found a novel way of gaining extra competitive leverage. It won a royal charter from the British government, creating a trade protectorate reserved exclusively to the company. The idea appealed to London, which wanted to block the region to the French and other rival European governments but – like many invading powers down the years – was too overstretched to do the job properly itself. According to one history of the time, the British vice-consul of the Delta's Oil Rivers Protectorate candidly admitted that London's policy would 'chiefly assume a negative character. So long as we keep other European nations out, we need not be in a hurry to go in.'

The charter awarded the conglomerate almost unlimited powers. It was subject to few controls and could effectively rule the Delta with the sanction and military support of the British government. Sir George Goldie, the UAC's head, highlighted the business's enhanced prestige by renaming it the Royal Niger Company. Its political connections were evident in the appointment of Lord Aberdare, a former home secretary, as its first governor. It was one of the great imperial multinationals of an era in which the East India Company and its peers held considerable sway in world trade.

The conglomorate's modus operandi was starkly laid out in the documents held in Britain's National Archives. Perhaps the most striking of them was a thick Foreign Office book so old and worn that it was bound together with a cloth tie. The yellow leather cover was fraying at the corners, exposing layers of hard cardboard beneath. On the spine, written in a burgundy-coloured box, was the title: 'Royal Niger Company Treaties. Part One 1891–98'. Inside, there was a covering letter by Goldie to the undersecretary of state for colonial affairs.

Writing under the ornate Royal Niger Company letterhead, Goldie introduced what he said was a full list of the treaties the company had signed. The following pages were divided by vertical red lines into three columns dedicated to the numbers, places and dates of the deals made by this capitalist buccaneer. The document had the air of a proud pupil seeking congratulations from his teacher for his industry: in one 20-day period in July 1889 alone, the company recorded that it struck six separate deals. According to British government records, the Royal Niger Company made between 340 and 500 of these agreements in total, to take spurious legal control over many territories of the River Niger.

The deeds were written in English and were mostly 'signed' by 'x' marks purported to have been placed there by local chiefs. The language in Goldie's book gave a sense of the scope of these agreements and how skewed they were in favour of him and his men. 'With the consent of our people and with the view of bettering their condition,' chiefs are quoted as agreeing in one deal, '[we] do this day cede to the company, and to their assigns, for ever, the whole of our territory.'

Armed with these documents, the company started trying to dislodge the palm-oil traders who stood between it and the commercial monopoly it wanted. Opponents were driven out, if necessary by force. The company soon established the dominance it desired: palm oil flowed to Britain and elsewhere in Europe, where it was used in making soap, candles and the explosive nitro-glycerine. Most importantly of all, it served as a lubricant, to stop the machines that were driving Europe's industrial prosperity from wearing out.

It was no surprise when the Niger Deltans sidelined in their own homeland bit back. A raiding party from Brass pulled down the Royal Niger Company's trading post in Akassa, killing several company officials. Britain responded by launching a naval attack on Brass, causing great damage and heavy – if still disputed – casualties. It also made plain that London planned to use ruthless measures to preserve its access to the Delta.

Over time, the attack would come to seem a defining moment in the Delta's history. The first hints of its wide repercussions are documented in the National Archives, in a report ordered by Britain itself. It was written by Sir John Kirk, a foreign office official, who on 11 May 1895 sailed from England for the Niger coast. His mission, as outlined in his later submission to the foreign secretary, the Marquess of Salisbury, was to carry out an enquiry into what he described in loaded terms as 'the recent outrage committed by the people of Brass'.

At first glance, the report looked as if it would be little more than self-justifying colonial whitewash. It was steeped in a presumption of basic British honour and an offhand contempt for the local people and their 'miserable' surroundings. The author thanked William Wallace, agent-general for the Royal Niger Company, for giving him 'every help and assistance' in his investigations. The work of company officials and British soldiers had, said Kirk, prevented him being put to 'considerable personal inconvenience'.

But, as Kirk's account continued, I could sense the doubts creeping into his own mind. He noted with disquiet that there was no official interpreter at the company's operations around Brass. 'In the whole service there is not an officer who can speak the native language,' he observed, 'which fully accounts for the want of knowledge possessed by the officials of what passes among the people.'

Kirk seemed equally troubled by what he saw of the trading rules the company had imposed, which were at the centre of the dispute with the people of Brass. Under its Royal Charter, the corporation was not supposed to run a monopoly, but Kirk found that in practice this was exactly what it did. Local people had to pay £150 annually to trade, terms that were, he said, 'impossible', because they 'practically exclude them from their old markets on the Niger'. Kirk was further disturbed by a visit to Nembe and Fishtown. He was struck by the 'degraded type of features' of many of those who had come back to rebuild their ruined homes. British troops had 'partially destroyed' Nembe and burned Fishtown to the ground, he found. In Nembe alone, he reckoned, 25 people were killed.

Kirk's reports of the Brass leaders' plaintive comments to him were equally disturbing. Although it was not clear how the two sides communicated, the Brass chiefs' testimony has a measured eloquence even when filtered through the twin barriers of language and a pen guided by a sense of British imperial superiority. They complained that the company prevented them from trading and killed many people each year. 'We do not want, nor expect, certain markets entirely to ourselves,' the report quoted them as saying. 'All we now ask is only to be allowed to trade at those markets where we and our fathers used to trade before the charter was granted to the Company. We are quite ready and willing to trade alongside white men.'

The company had done the local population 'many injuries', Sir John reported the chiefs as saying. It had barred people from other villages from coming to Brass to pay their debts to the traders there. Company officials had threatened to burn down other communities who did business with the Brass residents. 'We have frequently asked the Consuls that have been put over us . . . to tell us in what way we have offended the Queen to cause her to send this trouble on us,' Sir John quoted the Brass leaders as saying. 'The soil of our country is too poor to cultivate sufficient food for all our people, and so if we do not trade and get food from other tribes we shall suffer great want and misery.'

The Brass leaders went on to accuse the company of taking some of the abusive perks seized by any occupying force that has no fear of punishment. Kirk's report contained testimony from a woman called Akure, who said she was called up one day from her canoe to a Royal Niger Company hulk. Pregnant at the time, she obeyed the instruction out of fear. On board, she was raped by a clerk, who then offered her a present of a biscuit and two heads of tobacco. She refused to accept them, but was forced to take them on pain of being killed. She threw the presents over the side of her canoe. Later, she had a miscarriage.

For London, there were sufficient warnings in the report to suggest that the situation in the Niger Delta could become a big political embarrassment. Having congratulated itself for its pioneering role in ending the slave trade almost a century earlier, Britain now found itself accused of holding an entire region in bondage. If the story gained wide exposure internationally, Britain's commercial rivals could exploit its hypocrisy. Something had to be done.

In 1900, the British government revoked the Royal Niger Company's charter and itself took control of the Niger Delta. It was a change London made without much enthusiasm and at a great price. In a deal spookily echoed in twenty-first-century bank bail-outs, the Treasury immediately took on £250,000 of company debt, for which it paid face value plus 5 per cent interest. It also agreed to pay the company more than half a million pounds to cover territorial rights, disruption to its business and investment in development and equipment. In a final sting, the company was also given the potentially lucrative rights to half the royalties from mineral extraction from the Delta over the following 99 years.

The days of the charter might have ended, but the influence of the Royal Niger Company lived on. Its operations had established the Delta as a region where a system of militarized capitalism was used to further the interests of British colonization and commerce. Lord Salisbury, now prime minister, told Parliament that it was impossible to mention the names of men like the corporation's Lord Aberdare and Sir George Goldie without recognizing 'their great enterprise and resource, which place them high on the list of the pioneers of English civilization in the dark places of the earth'.

The Royal Niger Company's Delta putsch was part of a wider power-grab by the British in what was to become Nigeria. Having tried and failed to take the Islamic north by military force, London settled instead on a system of indirect government through favoured traditional leaders. In the south, British raiding parties scattered communities according to the doctrine of Joseph Chamberlain, the colonial secretary, who said it was necessary to use force to 'destroy the practices of barbarism, of slavery, of superstition which for centuries have desolated the interior of Africa'. As the historian Michael

Crowder has noted, the British found it hard to conquer southern Nigeria's clusters of decentralized peoples, who often had cultural affinities but politically 'were fiercely independent of each other'.

By 1914, Britain had decided that the vast territory it now controlled should become one, for economic and administrative convenience. It planned to fuse the northern and southern halves of the country, creating the state in roughly its modern-day form. The new country would be called 'Nigeria'. Lord Lugard, then governor-general, credited the name to his wife, Flora, a former colonial correspondent for *The Times* of London.

So Africa's giant was finally born, in circumstances all too familiar to many Africans, as a country that its people did not desire or design. It would be one of the most diverse lands on the continent or elsewhere, with several hundred ethnic groups speaking a similar number of languages. Predominantly Muslim in the north, its south was becoming increasingly – although by no means exclusively – Christian, as both Anglican and Catholic missionaries made inroads. The world religions were a patina overlaid on more ancient traditions of animism and other spiritual beliefs: the bearing of twins brought joy to the western Yoruba, but dismay to Igbo people in the east.

In one of those curious twists of historical fate, the creation of Nigeria coincided almost exactly with awakening interest in the country's potential as a source of oil. The hunt for crude had become a British strategic priority after Winston Churchill, the first lord of the Admiralty, had decided in 1913 to change the fuel of Britain's battleship fleet from coal to oil. This marked the start of a century shaped in good part by the geopolitics of crude. Churchill himself saw early on the profound significance of his decision. As he later described it, 'mastery itself was the prize of the venture'.

When Nigeria came into being in 1914, London declared that only British companies would have the right to prospect for oil. In 1938, with Britain approaching a world war in which oil supplies would be crucial, the government awarded a Nigerian oil exploration licence to a consortium comprising Shell and the Anglo-Iranian Oil Company, the forerunner of BP. The deal was the foundation of Shell's Niger

Delta operations today. Thus a colonial diktat launched an industry that would shape a nation and offer some of the most rewarding business opportunities in Shell's history.

Britain was now engaged in a cynical double-play in this largest of African imperial possessions. While it extolled the commercial possibilities of the Delta, it acknowledged privately that the wider nation it had created was doomed to dangerous instability. In short, Nigeria was a country to be squeezed for profit with little regard for its long-term future.

London's pessimism was clear as early as 1939, a mere quarter-century after Nigeria was created. A memo written by an unnamed official that year – and now lodged in the British National Archives – posed the question of 'whither Nigeria' and answered it in resoundingly negative terms. It was a document written according to the colonialist's grand intellectual sweep, with subjugated peoples serving as intellectual playthings for privileged minds to test out their theories of social organization, politics and economics. The official's conclusions – made 'with much diffidence, having no first-hand knowledge of the country' – would look to many modern Nigerian eyes full of justifiably bleak messages. He warned of the dangers of any attempt to 'pour the constitutional future of a vast country into a single mould'. If the metal was either 'inadequate or not sufficiently molten', he said, the result could be 'disastrous'.

The author suggested that the only way forward might be to go back to what had existed before. Nigeria, fused together out of imperial convenience a quarter of a century previously, should perhaps be decentralized, with each region becoming 'a practically self-governing institution', capable of managing its own affairs, but 'forming an integral part of a federation of many common interests'. In his conclusion, the author slipped from English to French. It might be a case of '*reculer pour mieux sauter*', he said: that is, of 'retreating in order to better make the leap'.

After the Second World War – and the dismemberment of the British empire that followed – London's worries about Nigeria grew more intense. In a July 1958 archive document, Sir Ralph Grey, deputy governor-general, recorded receiving a disturbing visit from Sir Tafawa Balewa, the country's prime minister. According to Sir Ralph, Sir Tafawa, a northerner, had confided that he knew Nigeria was not ready for independence and was going to 'have trouble' after it. His conclusion – 'He who controls the Police and the Army has the power' – has been underlined by a British official, as if in anticipation of the decades of dictatorship to come.

Sir Tafawa's sense of approaching menace was shared by British officials in Nigeria and Whitehall. A September 1958 memo from M.G. Smith, an official in Lagos, painted an unambiguously alarming picture of the country's prospects. It said the federal government had not developed 'strength and prestige' because ministers were 'unworthy of high office' and were 'always looking back over their shoulders' to maintain their positions in their regions of origin. Smith's conclusion had the fatalistic starkness of someone who knows they are talking about events now well beyond their control. He wrote, 'We had and still have considerable uneasiness about the ability of Nigeria, with its powerful and often bitterly opposed regional groups, to take the strain of independence as early as 1960.'

By 1957, after Ghana became the first official British colony in sub-Saharan Africa to win independence, it was clear that Nigeria's freedom could not be long delayed. The politics of the run-up to London's withdrawal were complex and are still much disputed, but many independent critics say Britain's actions helped to increase divisions in the new nation rather than reduce them. In an attempt to head off secessionist demands by northerners, for example, London introduced a new constitution that gave the regions greater autonomy. The effect – probably as intended – was to undermine the country's nascent nationalist movement.

There were straightforward political and commercial reasons why Britain would have wanted to do this. The leader of newly independent Ghana, Kwame Nkrumah, had already proved a strong, eloquent and charismatic critic of the West, threatening British interests in the region. The last thing London wanted in Nigeria was an Nkrumah-style nationalist and pan-Africanist leader. One in-depth piece of research on the Nigeria papers in the National Archives found 'repeated references in the late 1950s to the need for Nigeria to provide a counterweight to the more radical vision of Africa's future emanating from Accra' (Ghana's capital).

As Nigeria stood nervously on the threshold of independence, its political problems were far from unique. Similar concerns about stability existed in other African countries, many of which subsequently lurched disastrously into coups and conflict. But Nigeria, with its tens of millions of people, hundreds of ethnic groups and newly discovered oil reserves, had in some ways the greatest trouble of all. In hindsight – and arguably at the time – conflict was inevitable. Nigeria's politics were now dangerously dominated by crude, ethnically defined blocs, rather than the patchwork of caliphates, empires and highly autonomous villages that had existed previously. As the Nigerian novelist Chimamanda Ngozi Adichie has put it, 'Nigeria was set up to fail. The only thing we Nigerians should take responsibility for, in my opinion, is the extent of that failure.'

On 1 October 1960, Africa's most populous nation, then Britain's largest imperial possession, became independent, just 46 years after its masters had created it. What followed will be grimly familiar to students of war, occupation and so-called 'nation-building'. Less than seven years after independence, this complex land, confected in haste and then swiftly abandoned, would be the site of one of the twentieth century's worst civil conflicts. Once again, British political interests – not least the country's emerging oil industry – would play a central and highly contentious part.

In Brass that day, a short hop by boat from Akassa, I found that London's early losses in Nigeria's rolling battle for oil were still respectfully acknowledged. Nestled by a roadside in the town, neither obliterated nor preserved, was a cemetery containing some of the graves of British sailors who died during fighting that followed the 1895 Brass raid on Akassa. Unfenced and unkempt, the plot was flanked by shops and houses and the discarded clutter of everyday life, including a toaster, a dustbin lid and an empty plastic water sachet. Scattered among the graves, whose tenants ranged from a district officer to a Liverpudlian teenager killed by blackwater fever, were monuments to members of the crew of the HMS *St George*. Charles Charles, a leading seaman, George John Taylor, a lieutenant, and William Sutton, a domestic, were honoured by shipmates who 'deeply regretted' their deaths.

One epitaph in particular seemed to sum up the spirit of Britain's short but influential colonial adventure in Nigeria. On the ground, half-obscured by the drifting ocean sands, lay a memorial to Henry Byrne Chatteris, an HMS *St George* crew member and former King's Scholar at Eton. He died on 8 March 1895, a week after Henry Russell, whose grave I had seen across the water in Akassa. The Latin inscription on Chatteris' memorial spoke crisply of the bellicosity and sense of entitlement that had already come to characterize London's lust for Niger Delta oil. '*Quo fas et gloria ducunt,*' it read, or 'Where duty and glory lead.'

3

MORE POPULAR THAN CHURCHILL

In Enugu, a city built around hills north-east of the Niger Delta, one house stood out from the Saturday morning sleepiness enveloping the Independence Layout residential district. A note daubed in red paint on its high perimeter wall warned casual visitors against approaching a 'security zone' that was accessible 'strictly by appointment'. Beyond the gate was a black Mercedes-Benz 600 limousine, parked like a watchful elder brother next to a child's pedal car. The Merc's distinctive customised numberplate – 'Igbo 1. Eze Ndi Igbo' – laid its owner's claim to leadership of eastern Nigeria's millions of Igbo people. It was a contentious assertion, as some say the Igbo are notable precisely because they don't go in for such regal hierarchies.

The car registration was a clue that the person I was about to meet was no stranger to either power or conflict. Inside his house, I was shown to a front room dominated on one side by two large traditional thrones covered in animal skins. Opposite them was a – to Western eyes – more conventional collection of chairs and sofas. Sitting there was a bearded, square-jawed and heavily built middle-aged man, crouched over a late morning meal. As he turned to greet me, I did a double-take, for he was the younger image of my host and his father, Chukwuemeka Odumegwu Ojukwu, a man who knew more than most about the worst battle to date in the international war for Nigerian oil.

Ojukwu was the leader of a secessionist rebellion that occupied the country's oilfields more than 40 years ago, during what was to become a devastating two-and-a-half year civil war. His campaign for an independent eastern state, named Biafra, captured the imagination of many in the West during its own period of 1960s radicalism and social unrest. The worldwide fascination turned to horror as the conflict ended in one of the greatest annihilations of humanity since the Second World War, with many on the eastern side dying of starvation after the government stopped supplies to the breakaway state. The pain and poignancy of that time, when a nation briefly flowered and then died, are brilliantly evoked in Chimamanda Ngozi Adichie's bestselling 2006 novel *Half of a Yellow Sun*.

Beneath the death and the tale of a people's dream deferred, the war was a defining episode in the unfolding conquest of Nigeria's crude. Throughout, London armed the military dictatorship, which it saw as the better bet to protect British access and the interests of the oil companies, Shell and BP. Prime Minister Harold Wilson staked his reputation on the defeat of Ojukwu and his rebels. During the fighting, Wilson wrote that London had been 'steadfast' in supporting the Nigerian government, resisting 'all suggestions in Parliament and the press for a change in policy, particularly in regard to arms supplies'. It didn't take great imagination to hear the echoes of another war in another oil-rich state almost four decades on, prosecuted by another electorally successful British Labour prime minister despite widespread domestic and international opposition to his actions.

When General Ojukwu appeared in the living room, I could see immediately how he had become a leader of men four decades before. He looked imposing physically, but even more striking was his facility with words, expressed in a clipped southern English accent. This quirk, he said, laughing, meant he was referred to as the 'most Anglophile of all Nigerians', talking of African nationalism in the timbre of the landed Home Counties gent. He was well into his eighth decade by the time we met, with failing eyesight, but his manner still had a canine alertness. He leant towards me and spoke with the self-deprecation that only someone totally sure of his command could manage. 'Like a little lamb, I move forward to the slaughter,' he said. 'Relish this moment – you are the boss.'

I began by reading a few messages I'd been given the previous day by some civil war veterans I'd met. These wheelchair-bound men were something of a local institution, a living monument to a conflict whose legacy still looms large in Nigerian life. They sheltered in a lean-to by the main road west from Enugu, under rafters stuffed with T-shirts, condensed milk tins and newspapers. There they sat each day, talking politics and religion, discussing domestic problems and accepting any money offered by the sympathetic motorists who stopped from time to time. When I'd visited, they had spoken movingly of their hopes that the nation of Biafra would one day rise again.

As I read out their words about their enduring passion for Biafra, the general looked away, cocking his ear towards me. Then, slowly and carefully, he began to answer. 'There is no doubt what you have been listening to might be described as a bit melodramatic,' he said. 'But it is essentially the feeling in these parts.'

It was the start of a 20-minute monologue, a soliloquy worthy of Shakespeare. By turns passionate and reflective, the general was Prince Hal and Hamlet rolled into one. He barely hesitated or wasted a word as he talked with the measured deliberation of someone who was used to being listened to intently and without interruption. As theatre, it was spellbinding, the kind of speech that makes the hairs on the back of your neck stand on end. I could see within two minutes why an Igbo friend of mine described the general as more popular in eastern Nigeria than Churchill had been in wartime Britain.

General Ojukwu said he did not like to talk about the war now. It pained him, and he did not like to wallow in that feeling. Neither did he want to be made the symbol of a conflict whose story needed to be 'told from a million mouths'. He talked emotionally of the veterans crippled 'for the deepest, most valiant, most glorious reason a man can have'. 'They are in their wheelchairs today because they believed in their own people,' he said.

He listed what he saw as Biafra's achievements during the civil war: living without imports, maintaining 'perhaps the busiest airport on the continent of Africa', broadcasting its message to the world nightly from communications equipment in the back of a truck. He reminisced with the aide and former wartime comrade sitting next to him about how no beer had ever tasted better than the brew of Biafra. It was these things that had stuck in the minds of the Igbo people, he said, their lustre growing as Nigeria continued to be a grim place to live. Many people had seen Biafra as the first country in Africa to truly liberate itself. 'Those were the days,' he said, articulating in a wistful aside what he thought many Igbo people believed. He returned to his own voice, asking rhetorically, 'You say, how can those be the days?' before answering his own question. 'Those years we were independent. Those years we were allowed to grow.'

One of the most striking biographical facts about this unapologetic secessionist, so loved by so many of his people and so despised by so many of this enemies, was how many years he had spent out of his homeland. His 'problem', as he put it to me, was that, at the age of 12, he had left Nigeria to be schooled in England. The son of a transport magnate and millionaire, he was educated at Epsom College and Oxford University. He learned everything he knew from England, he said. 'But what else did I get from it?' he asked savagely. 'Pain!' he immediately replied, spitting out the word like over-chewed tobacco. He continued, 'Because I have taken the precepts that have made Britain and tried to transfer them to Nigeria. And all I got from Britain was betrayal.'

The long wind-up to Nigeria's civil war began in rising public discontent with the British-backed post-independence regime. In January 1966, the centre could no longer hold: a group of army officers, nicknamed the January Boys, launched a military coup, killing prime minister Sir Tafawa Balewa and other senior politicians. Horrible though the violence was, few people in Nigeria or London were under any illusions about the quality of the government that had just been overthrown. In a 1968 memo to Prime Minister Harold Wilson, a foreign office official in London said the coup was led by 'a group of young radicals in the army' who were dismayed by a 'thoroughly corrupt regime' that had superintended 'disgracefully rigged elections'. The takeover was greeted with 'jubilation', the official concluded, showing that ordinary people all over Nigeria were 'delighted by the change'.

The new master of Nigeria was Major-General Johnson Umunnakwe Aguiyi-Ironsi, who became emergency head of state. He planned, contentiously, to abolish the country's regions and introduce a unitary government. But powerful northerners denounced the action, portraying it as an attempt by Ironsi and his eastern Igbo kinsmen to take over the country and break it up.

In July 1966, the Ironsi regime was overthrown by another military coup, led by the northerner General Yakubu Gowon. In September, people from the east, whose entrepreneurial trading culture had scattered them across Nigeria, were attacked, particularly in northern cities such as Kano. The aggression became a pogrom, in which at least 10,000 easterners were slaughtered and many times that number fled to their homeland.

The sense of crisis deepened as General Gowon and his political opponents bargained over the country's constitution in talks in Ghana in January 1967. The negotiations failed after General Ojukwu, then the eastern region's military governor, claimed that General Gowon had gone back on agreements made. In May 1967, General Ojukwu declared the secession of the eastern Biafran Republic, with himself as head of state. General Gowon imposed an economic blockade on the east, and in July the first shots of the civil war were fired.

The conflict's outbreak revealed starkly the internal regional divisions entrenched in Nigeria by colonial Britain. It also placed the Labour government of Harold Wilson in a position of considerable responsibility. As the former colonial ruler and a regular supplier of arms to the Nigerian government, Britain would have a strong influence on the conflict's outcome.

From start to finish, London's position was one of unequivocal – and highly controversial – support for the Nigerian federal government. Arms continued to flow to the then federal capital, Lagos: London argued that it was merely continuing existing policy and fulfilling a duty to help defend a fellow Commonwealth member whose land had been invaded. No protests from Biafra, or its sympathizers overseas, changed what Britain maintained was a principled stance. Nigeria must remain whole and alliances must be honoured, London said, much to the Lagos authorities' relief and the Biafrans' dismay.

What Britain was more coy about saying in public was that one of its biggest preoccupations was protecting its stake in Nigeria's oil. Biafran forces had started the war strongly, occupying Port Harcourt and other oil-producing areas. This had a potential impact on both crude supplies to Britain and British commercial interests. Before the conflict, Shell and BP, operating in a joint venture, accounted for about two-thirds of the country's oil production of just over 1m barrels a day.

General Ojukwu's army moved to within 130 miles of Lagos, the capital, before government troops checked its advance. Then, helped by Britain's support, the military dictatorship began to turn the tide. Federal forces began a long siege of Biafran territory, recapturing Port Harcourt in 1968. By now the blockade of the remaining Biafran territory was causing devastating food shortages, mitigated only by relief flights run by agencies such as the Red Cross and the Catholic group Caritas. Non-governmental organizations and journalists were sending stories and pictures of starving Biafran children around the world, but the federal government kept up the blockade and bombed eastern settlements and farmland. Finally, in January 1970, the federal forces broke the heart of the resistance. General Ojukwu fled Nigeria, days before the surrender of his forces. He did not return for more than ten years.

No one knows how many people died in the civil war, although all agree that the numbers were huge. The lowest estimate of deaths is about half a million; the highest is six times that. The casualties dwarfed those in the Blair-era Sierra Leonean civil war, during which Britain – in contrast to its behaviour in Nigeria – deployed troops to help stop the fighting. The vast majority of deaths were on the Biafran side, as the economic blockade by the Nigerian government starved civilians as well as the military. The rebellion was crushed and the eastern region emasculated, allowing the business of drilling the Delta's oil to resume.

General Ojukwu told me he believed the question of oil was critical to understanding British policy during the civil war. Perhaps telling me what he thought I wanted to hear, he said he thought this was what London's strategy was all about. Once more, he was talking not as an outsider but as someone with more than a passing familiarity with the international politics of crude. His father had been a director of Shell just before the war, and the company used to fly the young

Chukwuemeka home from England for holidays to the eastern cities of Owerri and Port Harcourt. 'So to a large extent I grew up under the umbrella of Shell,' he laughed.

At the start of the war, he said, the Biafrans had looked to Britain – so recently the occupying power – as a sort of mentor. They had sought understanding for their cause and been shocked to find none. 'What was it you taught me?' he asked bitterly, before beginning a vivid recitation from the arch-imperialist Thomas Babington Macaulay's poem 'Horatius'. The poem is about the defence of a bridge by the eponymous Roman protagonist and two of his colleagues against the invading Etruscans. Written, according to the author's conceit, by a Roman citizen of the time, it is steeped in patriotism, and dismay at corruption and factionalism. The author is also, as one critic puts it, given to 'pining after good old times which had never really existed'. It is the kind of poem a 1940s Epsom College schoolboy like General Ojukwu would have learned, and it is the kind of poem the leader of the surrounded Republic of Biafra might be expected to recall.

In the two verses the general read, Horatius's intervention begins to turn certain defeat into possible victory:

> *But the Consul's brow was sad,*
> *And the Consul's speech was low,*
> *And darkly looked he at the wall,*
> *And darkly at the foe.*
> *'Their van will be upon us*
> *Before the bridge goes down;*
> *And if they once may win the bridge,*
> *What hope to save the town?'*
>
> *Then out spake brave Horatius,*
> *The Captain of the Gate:*
> *'To every man upon this earth*
> *Death cometh soon or late.*
> *And how can man die better*
> *Than facing fearful odds,*
> *For the ashes of his fathers,*
> *And the temples of his gods?'*

The general reverted to his normal speaking voice. 'That was the height of courage, the height of patriotism,' he said, adding witheringly, '*That* was what you taught me.'

General Ojukwu's dim view of London's conduct during the war was supported by some extraordinary official documents of the time that I'd seen during my visits to Britain's National Archives. Prime Minister Harold Wilson set the tone in a memo sent to Michael Stewart, his foreign secretary, three weeks after the start of the conflict. In it, Wilson said it was of 'national importance' to protect the Nigerian investments of Shell and BP, which were crucial to Britain's balance of payments position and post-Second World War economic recovery. Everything should be done 'to help Shell-BP and the federal Nigerian authorities to establish effective protection of our oil investments'. It was a position from which Britain never seriously wavered, despite years of conflict and many hundreds of thousands of deaths.

The archive files revealed the prolonged shared anxiety of Shell-BP and British government officials about the fate of the company's operations. An August 1967 memo to Wilson from George Thomas, minister of state for Commonwealth relations, claimed that Ojukwu had sequestrated Shell's assets and ordered the departure of its employees after it refused to pay royalties to the Biafran government. The letter ended with the observation that the oil industry had 'much to lose' if the Nigerian government did not achieve its 'expected victory'.

In private, Shell put considerable pressure on British officials to keep sending arms to the federal government, even though it was an unelected dictatorship. Some of the lobbying efforts were clearly described in a June 1968 memo written on the day of a parliamentary debate on Nigeria. A foreign office official noted that he had received a call from a Shell executive to say that the company felt that supplying arms to Nigeria was 'absolutely right'. Shell attached 'enormous importance to it' and feared for the consequences of a change of policy. In Shell's view, the official concluded, such a shift would 'greatly impede' the company's chances of re-establishing itself in Nigeria and resuming operations at Port Harcourt.

Shell, like the British government, had taken a huge bet on a federal win in the war and was anxious to do what it could to ensure that outcome was achieved. Unsurprisingly, the company was keen to keep its advocacy confidential. At the bottom of the letter, a hand-written note by another government official suggested this sneakiness was not lost on Whitehall. 'It would be better for Shell,' the note said, 'to tell the critics this and arrange some public support!' Shell had no comment when I asked if these documents gave an accurate account of its position at that time.

As the war ground on, there was no let-up in either Britain's supply of arms or its defence of the oil companies. In August 1969, the foreign secretary, Michael Stewart, fired off a reassuring response to an anxious memo from Wilson that again raised concerns about the threat the war posed to the industry. Stewart reported that, at Britain's suggestion, the Nigerian High Commission in London had just ordered 48 anti-aircraft guns, 60 machine guns and ammunition. The British government would 'do its best' to see that these weapons were used to protect Shell-BP oil installations. He gave Shell the names of five retired British military officers who would be willing to act as advisors on 'passive defence'. Stewart also planned to offer the Nigerian dictatorship a visit from an RAF air commodore and group captain to advise on the air war against Biafra, which had been condemned because of the number of civilian casualties it inflicted in markets, residential streets and other densely populated areas. The air force officers would look at whether there was anything Britain could do to help the Nigerian government combat Biafra's own warplanes and 'remove the Biafran air threat to the Shell installations'. The memo ended by expressing the hope that the British officers' visit to Nigeria would not attract press attention. 'We should certainly wish to avoid any impression,' Stewart wrote, 'that Her Majesty's Government were directly involved in the conduct of air operations in Nigeria.'

The documents I saw were candid about the ruthlessness with which Britain planned to protect and increase its profits from Nigeria's nascent oil industry. In one memo, an official assessed the commercial impact of the war on Britain. He spotted an advantage for London: war damage to Nigeria's refineries meant the country was having to import increasing quantities of refined petroleum products, most of which would probably be bought from Britain. 'These imports are costing Nigeria anything up to £5m a year,' he noted, concluding that 'the United Kingdom must be earning quite a substantial net profit'.

The official went on to list other reasons why London should not worry too much about people being killed by British weapons or starved to death under a blockade supported by military hardware supplied by London. On food supply, the Biafrans had rejected some Red Cross relief; more generally, West Africa had always been an area of chronic protein shortage and 'consequent malnutrition, semi-starvation and starvation'. The conflict itself was 'a very lackadaisical affair', in which troops knocked off for two or three hours in the middle of the day for a 'truce or siesta'. During action, most of the time they just 'shut their eyes' and fired their guns in the air, causing 'prodigious expenditure of ammunition and remarkably few casualties'. Throwing a final comfort blanket over any lingering moral concerns, the official said there had been only brief intervals of 'real fighting'. 'Most of the time,' he wrote, 'nothing whatever happens.'

The most frank insight of all into Britain's approach that I found was the June 1969 valedictory dispatch prepared by Sir David Hunt, the departing high commissioner to Nigeria. Written by a man who confessed to feeling he had spent 'long enough' in the country, it claimed London's policies had 'successfully defended British interests, restricted the increase of Russian influence and won the respect of the Nigerians'. Echoing the assessment of generations of colonial-era British officials, Hunt concluded that the need to safeguard British investment in Africa meant a united Nigeria was 'in the British interest'. Shell-BP was only just starting to make returns on its investments after 'long disappointment'. The decision to recognize the federal government and not Biafra 'defended our major commercial interests and kept us on the right side with the rest of Africa'. Even better, the Nigerian government had always paid cash for its weapons, 'to the incidental benefit of Britain's adverse balance of trade'.

Cant and hypocrisy were the 'besetting sins' of Africa, Hunt lamented, not pausing to consider whether Britain's own position could be viewed in the same way. 'Nor is Black Africa as far removed from barbarism as other continents,' he continued. 'It is not so long since it was the practice to enslave or eat such of your enemies as were not required by the religious authorities for sacrificial purposes.'

Hunt concluded that he had a 'great affection' for Nigerians, who deserved a 'happy and united future', even though their country was, regrettably, full of 'small tribes whose very names tax the memory'. Nigerians, he said, were generally 'cheerful and friendly', despite a 'maddening habit of always choosing the course of action which will do the maximum damage to their own interests'. At least Nigerians usually made their blunders 'with an engaging air'. Politics aside, the Biafrans turned out to be the Nigerians he liked best because they were 'the most intelligent and the most hard working.' What was implied between the lines was that these resourceful people should know their place.

General Ojukwu was a little evasive when I asked if he, a highly educated man, seriously believed that Britain's actions in an international crisis replete with oil interests would be consistent with its avowed values of individual liberty, decency and human rights. Nigeria, after all, is one of many parts of the world whose history suggests otherwise. The general's reply was insouciant and carried more than a hint of false modesty. 'If you say I was naïve, you can be right,' he said. 'I am not that brilliant.'

He insisted he never really saw the British system and its political machinations from the inside. He said, self-deprecatingly, that he'd had a 'certain aspiration to join the gentry' when he was at Epsom College. But he'd always remained an outsider. For all his years at public school and Oxford University, he was never really integrated into Britain and its power structures. 'You don't see much,' he said. 'You don't.'

The general did concede that the politics of the war weren't easy for Britain, but he argued London should have been 'more understanding'. The British had seen the deterioration after independence in the country they had created. 'It wasn't really a surprise to anybody,'

Ojukwu said. 'So I ask myself, if it wasn't really a surprise, was it welcome? Or, if it wasn't welcome, what did you do to stop it?'

Remarkably, one of the worst wars Africa has ever seen ended without the reprisal massacres of easterners that had been widely feared. Both sides were tired after the long attrition. The Biafrans had many supporters overseas and had proved far better than the federal government at evoking a sense – whether justified or not – that they were pursuing a noble cause. In a famous radio broadcast, military dictator General Gowon declared that the war had produced 'no victor, no vanquished'. The resentments that had led to the conflict would burn on, but for now Nigerians demonstrated what many Africans have had to become so good at: a kind of pragmatism under duress, an acknowledgement that life must go on. As a British journalist who covered the conflict put it, 'There was chemistry arising from the evolution of the war that somehow seemed to diminish the lust for revenge.'

For postwar Nigeria, uneasy, unhappy reconciliation was the only viable option. At least the nation had the advantage of entering a new phase of booming oil development that promised to help it buck the African trend of mass poverty and infrastructural decay. Yet, a little more than a decade later, that very same wealth from crude helped push it to the brink of one of the swiftest and most dramatic economic collapses suffered on the continent or elsewhere. Unsurprisingly, outsiders – including Britain – were at the heart of the action, stoking a conflict over oil wealth that continues to this day.

Nigerian oil boomed after the civil war because of the combination of fast-rising production and the worldwide price spike that resulted from the Middle East oil export embargo. In 1964, Nigeria's crude oil exports earned, literally, less than peanuts; seven years later, they had increased 15-fold and accounted for 79 per cent of export earnings. It was the start of an era in which, economically, oil was just about the only game in town.

Everyone – at home and abroad – wanted a seat at the feast. In 1969 – with the civil war still going on – Britain's foreign office began a long-running and wide-ranging debate about how London's commercial interests could be protected and promoted in a world of cascading cash and dodgy deals. Newly oil-rich Nigeria, an official observed at the outset, formed the '*locus classicus*' in West Africa for 'post-independence corruption and for the system of "kleptocracy"'. For all the need to 'clearly point out to British firms the long-term dangers of encouraging graft even if the short-term advantages seem attractive', there were realpolitik considerations to be borne in mind, too. British companies shouldn't automatically dismiss the idea of paying bribes; better instead to take advice from old Nigeria commercial hands about 'how far they should play the Romans in Rome'.

Another official, writing after the civil war ended, mused about what lessons London could learn from France's success in doing business in Nigeria. Basically, he said, the Nigerians never allowed politics to 'come between them and their commercial interests, whether public or private'. A good deal of political mistrust still existed between Paris and Lagos – not least because France had supported Biafra – but this was 'pushed aside' when there was 'any question of turning an honest penny or two'.

The memo continued in the same tone of innuendo, grudging admiration and low-class irony. The official listed French skills and advantages as 'confidence', 'bribery' and 'uncompromising policies', accompanied sometimes by a 'bland admission of guilt', but never by 'any disposition to wallow in it'. All this appealed more to the Nigerians than 'what they see as the hypocrisy of other nations', he noted, 'including, I fear, our own'. British policy should, he concluded, take 'a tougher, less scrupulous line with, always, an eye to the main chance'.

This approach would serve London well once the return of civilian government in 1979 sparked the start of a spending and borrowing boom that would eventually turn into a Nigerian version of the subprime lending crisis. In January 1981, the new administration of President Shehu Shagari and vice-president Alex Ekwueme began a five-year plan for a huge, mainly oil-funded, $154.8bn investment in public and private projects. Britain's view of these developments was summed up bluntly in an internal memo on trade that was circulated at the British High Commission in Lagos. It was entitled simply, 'We need to grab our share'.

The impact was immediate. British exports exploded into life, doubling in a year. Between January and October 1981, the freight carried by the state-owned Nigeria Airways from Britain to Nigeria was more than 200 per cent up on the same period of the previous year. It is hard today to imagine this freewheeling atmosphere of business bonanza, in which Nigeria was eagerly courted by foreign industrialists. The idea of an African country having money and grand plans will look ridiculous to many Westerners accustomed to seeing such nations through the 'Live 8' lens of unremitting poverty. Yet those who know those times well depict them as a feeding frenzy of consumption. A Whitehall official once described to me how thousands of British companies were involved in the supply of products ranging from 'widgets to fridges', as well as 'bigger, meatier stuff', such as hotels, utilities infrastructure and weapons. The business between the UK and Nigeria was so 'intense and varied', he said, that it covered almost 'every conceivable' trade sector.

Yet – just as during the birth of Nigeria during colonial times, or the financial crisis now – ill omens were there at the time for those who cared to look. Almost from the start, the stories of investment were interwoven in the press with tales of failed projects, overspending and corruption. Even as Nigerians and foreigners were galloping Nigeria on into the industrial age, it was clear the country was never going to make the jump.

Richard Synge, then editor of *Africa Economic Digest*, a London-based magazine, characterizes the era as Greek tragedy. 'You could see it coming,' he said, 'but you didn't know how it was going to unfold.'

By the end of 1982, Synge's magazine was noting that the optimism expressed by foreign bankers and businessmen about Nigeria in 1980 and 1981 had given way to 'unrelenting gloom.' Oil revenues had fallen 42 per cent in two years as the price of crude collapsed from $40.97 a barrel in December 1980 to $28.25 in March 1982. In a sure sign of a country in crisis, the government turned its hostility on immigrants, which it expelled en masse. The large Ghanaian community was hit particularly hard: a quarter of a century later, the ubiquitous large checked holdalls that many West

Africans use when travelling are known universally in Nigeria as Ghana-Must-Go bags.

By December 1983, Nigeria was in crisis talks in London over its unpaid debts. British banks estimated that Nigeria's foreign debt had ballooned from between $4bn and $5bn in 1981 to $10bn in 1982. It was a burden, swelling each year with interest, which Nigeria was to be saddled with for almost a quarter of a century. Freedom of a sort came only when the country received $18bn of debt relief from the Paris Club of creditors in 2005, in exchange for a $12bn partial payment in return.

Britain has never accepted responsibility for its part in creating a problem that haunted Nigeria for more than two decades. An official once argued to me – as governments have recently claimed about the international financial crisis – that nobody else spotted the looming problems either. Britain, he said, had no need to reproach itself or cancel debts that many felt it had recklessly helped generate. 'I don't think it's the case,' the official told me, 'that we would write off debt because we had some sort of guilt, conscience, or because we felt we had underwritten debt that we shouldn't have underwritten.'

It's not an argument that carries much weight with Alex Ekwueme, vice-president during those turbulent financial times. I went to see him because he had a reputation as one of the few honest people in the civilian government. From time to time since, people have called for him go back into politics, but he has never managed to make headway. It seems likely that part of the reason why some Nigerians want him in high office is that he is rather ambivalent about the idea itself: even if he craves power, he does not seem to desire it nearly as badly as some of his rivals. He travels with the standard siren-serenaded official convoy, but, unlike many other politicians, doesn't give the impression particularly of relishing it.

When I met Ekwueme, at his house in his home town of Oko in the east of the country, it was hard to picture him as the deputy head of a spendthrift, irresponsible and corrupt government made drunk by oil riches. Everything about his appearance exuded an air of precisely the formality and control that his administration was said to so conspicuously lack. He wore a dark suit, red tie, white shirt, tie-clip and hexagonal cuff-links. He had a neat scrub of white hair and a clipped matching moustache. As he spoke, he kept his legs folded, his right hand resting on his thigh and his left clutching an armrest. He made only economical gestures with his hands and spoke quietly, at times to the point of mumbling. In almost every respect he seemed the antithesis of the grandstanding politician and boastful big man; it is perhaps a clue as to why his time in politics proved such an unhappy experience.

'I didn't set out to be vice-president,' he said diffidently. 'It just worked out that way.'

A businessman with interests ranging from brewing to architecture, Ekwueme had a reputation as a philanthropist, having funded projects from a local school to the huge, half-built church just down the road from him. His house was of unostentatious brick and concrete construction, with a neat flower garden and no razor wire on the wall. Inside, the political memorabilia included a leopard skin and an elephant tusk, both official gifts he had received as vice-president, in times before such largesse was circumscribed by international environmental laws. On top of the television was a trophy shaped like a map of the country. It bore a debatable inscription: 'The architect of Nigeria's stability'.

Ekwueme said his government had struggled from the start because it took over in tough and unusual circumstances, after a voluntary handover by the then military dictator General Olusegun Obasanjo. It was the first time the army had relinquished power in a country in Africa, where coups and counter-coups had become common. It meant an inexperienced elected government had taken power after a disputed election in a country that was still trying to build institutions after the privations of colonialism, dictatorship and a devastating civil war.

To add to the dangerous mix, world oil prices had surged from $14.05 a barrel in December 1978 to $38.69 a year later, bringing cash flowing in and encouraging dreams of industrial prosperity. Investment plans were drawn up by economists who bargained on the oil price staying high, Ekwueme said. International banks and other foreign companies were only too pleased to offer loans and imports to satisfy the new demand from Nigeria. Because Nigeria had visible income, Ekwueme said bitterly, 'the sharks, the lenders' were 'falling over each other' to sell products to Nigeria that eventually left the country with little to show other than a mountain of debt.

If Ekwueme wasn't exactly trying to absolve himself of all responsibility for this, he certainly wanted to spread the blame around. He referred pointedly to the 1970s decision of General Olusegun Obasanjo – who later re-emerged as a civilian president who enjoyed far more popularity abroad than at home – to take out Nigeria's first 'jumbo' loan from foreign banks. Ekwueme's position seemed to reflect *Africa Economic Digest* editor Richard Synge's observation that 'everyone and no one' was culpable in the oil boom débâcle. As with our current economic turmoil, what happened reflected the frenetic spirit of the times.

In this conversation about a crisis almost three decades past, this quiet, understated, man was at his most passionate when talking about the role of Western governments in the collapse of Nigeria's oil bubble. He did not subscribe to the view once advanced to me by James Wolfsensohn, the former World Bank president, that foreign institutions bore at most 10 per cent of the responsibility for Nigeria's debt crisis. Ekwueme said they should have known which of their companies did not have good track records and were taking money from Nigeria for shoddy work. He talked about a paper mill the government decided to build because a study showed it would be profitably run and would generate enough income to pay off the loan needed for the construction. That turned out to be a false premise, leaving Nigeria with a debt that could not be paid once the oil price collapsed.

It was a period of boom and bust that set Nigeria on the chaotic path it has careered down ever since. By impoverishing the many at the expense of a corrupt few, it helped set the stage for the prolongation and expansion of the country's internal war over crude. The turbulent life of modern Nigeria and the scores of millions of very poor people trying to scavenge within it had been determined. From now on, the nation would be an emblem of misrule, its inhabitants trapped in an oil-dependent country that – as one journalist has put it – was like a bad marriage no one wanted to leave for fear of losing a share of the spoils.

Ekwueme said that he thought rich nations, including Britain, were far more culpable than they had ever acknowledged. London and its fellow capitals in the Paris Club of international creditors expected and received plaudits in 2005 for accepting a partial payment of Nigeria's loans and cancelling the rest. But, looking back at the crisis that had generated this vast debt more than 20 years before, the former vice-president took another view. If debt relief negotiations between Nigeria and the West were ever cast in moral terms, he said, 'we shouldn't be talking at all. They should be giving us some change.'

PART TWO
LIVING AT THE OIL FRONTLINE
Why Lagos is Nearer London than You Think

'Monkey work, baboon chop.'
    – popular West African aphorism

4

THE BOYS FROM THE BOOKSHOP

At the heart of Lagos Island, a statue of the nationalist hero Herbert Macaulay bears silent witness to a great cheer that has just erupted around him. It's coming from a group of young men in the courtyard of the dilapidated public library opposite, whose perimeter metal bars make it look more prison than place of learning. A broken window reveals a scraggy and uninviting book collection, but the men are in any case looking elsewhere. They are greeting their boss, Adekunle Godwin Talabi, who walks in front of them with the cool and cockiness of Rocky entering a boxing ring. Talabi gives a little wave of acknowledgement to his support chorus, which continues to howl with a volume worthy of the Las Vegas cheap seats.

The serenade demonstrates the power of Talabi, one of a new generation of bosses in a society where Macaulay once held sway. In this small slice of a great metropolis – the former capital and still the country's biggest city – Talabi is the sovereign master of government and business. He combines the roles, among others, of tycoon, police force and immigration officer. His commercial interests include DVD retail, a barber's salon, a building contractor and a security company. Most crucially, he is the 'chief of all boys'. 'I am the chairman,' he says. 'I am the manager.'

This means he is the head of a group of young – and not so young – men who are known universally as 'area boys'. These gangs control large parts of Lagos Island, the city's historic centre, extracting money from outsiders and insiders who want to do business there or simply to pass through. Talabi – whose modest but apparently undisputed domain runs from the nearby Bookshop House crossroads to the northern end of Odunlami Street – describes his boys as his 'army staff' and 'soldiers'. He compares – with just a touch of melodrama – the situation of oil-ravaged Nigeria to that of Liberia during its notorious civil war, when militias roamed, raped and robbed their way around a fragmented nation. 'Nobody can say what's going to happen,' he says. 'So everyone is fighting for his own life, his own living.'

Talabi's men are, in their way, as emblematic of the distorting, destructive influence of Nigeria's oil economy as Asari's Delta militia men. The Lagos gang fight in the no-man's-land between the breakdown of traditional power structures and the rise of global capitalism. The group has no elders, Talabi says, and no wider sense of patriotic loyalty. They see themselves – or so, at least, they tell me – as a symptom of lawlessness rather than a cause of it. 'I am providing security for my area, not for my country,' Talabi says. 'For my area, for my people – in order not to let us fall into bad temptation.'

Urban gangs are hardly unique to Nigeria, but what makes Lagos's young male scavengers striking is that their roots lie as much in the limited riches brought by oil as in grinding deprivation. When wealth spills from the ground, it makes sense to try to get one's hands on a piece of it rather than to make some productive business work and then take a cut from that. In other highly corrupt states, the criminals know that they have to give the real economy space to flourish if they are to take their share. But if your nation's wealth is extracted by hired hands, such as the international oil companies, you need do nothing more than snatch your slice. The area boys are part of a grab-it-now free-for-all that helps explain Nigeria's paradox of poverty: its oil makes it richer than most African countries, yet the scale and scope of its social decay rank in some ways among the worst on the continent.

Like the other hustlers who flourish in Lagos's myriad social crevices, the area boys are a reminder of how easy it is for a help-yourself culture to become ingrained in societies where institutions have failed. Talabi's gang and its like are not the atavistic 'feral youths' of Western bourgeois nightmare, but pragmatists playing a broken system. Nigeria's oil – and the income it yields every day – has simply magnified and made visible threats that face many societies in both the rich and poor worlds because of the way most of us live now. Nigeria's dependence on a single, volatile source of income has echoes in Western countries that have grown increasingly – and, we now see, dangerously – reliant on the apparent ingenuity of the finance industry and the invisible movement of money. If the world's leading retail banks ever did fail – even temporarily – how long would it be before cash shortages and the inability to buy food brought area boys and girls of all classes to loot on London's streets?

I had wanted to spend time with Lagos's area boys ever since an encounter I'd had with some of them during my first weeks in Nigeria. One Saturday, a friend and I were approached by three youths as we were having a look round Lagos's Balogun market area. Their introduction was to the point. 'We are your friends in this area,' said one. 'Please empty your pockets.'

Through arrogance or naïveté – or both – we refused and carried on walking. The young men insisted we pay them. We held out. A debate began and continued surreally all the way back to our car, still unresolved. We ended up exchanging goodbyes with them and driving off, intrigued and puzzled by perhaps the world's most courteous attempted mugging. It was almost as if the young men were anxious for some kind of validation and acquiescence on our part, so that what they were doing seemed less like extortion and more like a paying of dues and observance of custom.

From that day, I was struck by the way area boys flitted in and out of my life and work in Nigeria. Their presence – part mischievous, part menacing – was the inevitable accompaniment to many street transactions, from buying palm wine at a bus station to gaining entry to a beach on a Sunday. When my parents visited me, we found our intended route on a drive through Lagos blocked by some young men holding a chain across the road. When we diverted to another street, they made no attempt to stop us. They could have been tax collectors, or robbers – or both.

On first meeting, Talabi looks more fiscal bureaucrat than hardened criminal. He doesn't seem to have much in common with the drugged, abused, cross-dressing child combatants who ruled the Liberian conflict he evokes. He's courteous, coherent and dressed unextravagantly in a blue French football shirt. His good looks have survived the sizeable scars he has sustained above his right eye and right ear. It all makes him an unlikely, roguishly charming and rather superannuated 'young soldier', typical of the middle-aged Nigerian men who, infantilized by lack of work, routinely describe themselves as 'youths' or 'boys'. In Africa's most populous state, it often seemed to me that the seven ages of man had contracted to two: unemployed and retired.

Like other members of Lagos's thriving hustling class, Talabi is a protean fellow, hard to pigeonhole in terms of upbringing or behaviour. One of the first things I notice about him is that he is far from the bottom of the social pile in terms of background and opportunities. A forty-something father of two, he describes himself as a 'well-educated somebody'. He says he went to school but had to forgo university because of financial difficulties: his father had left his mother, leaving him to look after his younger brother. 'I can read, and write, and behave the way you are going to like,' he says, disarmingly.

The territory his gang occupies is a living history of a city founded on imperialism and oil-driven commerce and – ultimately – hobbled by the hangover from both. At one end are the banks and the key colonial highway of Broad Street. At the other, past coffin shops and a rank pile of roadside rubbish, is the social centre of Campos Square and its attractive but decayed houses, built by the returnee descendants of slaves who had been taken to Brazil. It is, in Talabi's lyrical evocation, 'a golden unit, a golden area', in spirit if not in modern-day appearance.

Just like Asari's gang, the Bookshop House boys claim they provide their area with security that the authorities are incapable of delivering. Talabi's group say they don't extort cash, although they are always willing to accept donations. Money is certainly one of their key motivations: as Talabi talks about his 'fight for the youth to grow in order to make them rich', one of his men echoes his words as if in a cappella recital. The men are also 'regulating traffic' in the area, in their capacity as officials of the National Union of Road Transport

Workers, but they insist that they do not take money from drivers. Talabi himself runs three minibuses – yet another business in his bulging portfolio.

As Talabi talks, some of the boys from the library begin to gather round and listen. For the purpose of the interview, their boss has commandeered a street stall where people pay to make mobile phone calls. The owner has acquiesced graciously – she had little option – and gone off to eat some stew and rice. Congregating around the counter, the men are eager to talk about what they are doing. Kwam Anthony, one of Talabi's deputies, lists his own business interests, speaking as if reciting a prayer. Like his boss, he is a father of two children.

The boys say much of the gang's earnings are set aside and put towards getting its members out of Nigeria to countries where there may be better opportunities. Anthony says that two boys have left Nigeria to start a new life in Spain, where they are working as driver and window cleaner to a wealthy Nigerian. Two more are due to go the following month. Talabi himself is in fact only the vice-chairman of the group; the chairman is in London, trying to find work as a security guard. The gang want to formalize their travel-agency arrangements, registering themselves as a company called 'Young Shall Grow Ventures.' The idea seems a curious mixture of street smartness and sentiment, articulated by petty criminals and hoods who seem to yearn for the kind of love, respect and comradeship that they often don't show for others. 'We are going to help each other,' Anthony shouts, overcome by a wave of feeling. 'In order to make everything grow.'

The boys are Christians, who see no contradiction between their beliefs and their everyday currency of extortion and intimidation. Just as in Asari's militia, religion is put to work in the service of the cause. At one point, Talabi holds an impromptu prayer to appeal for 'strong power'. As the boys chorus 'Jesus Amen' three times, he intones, 'All my enemies are falling in the name of Jesus.'

Suddenly, a young man not from the group materializes and starts berating Talabi. The chief of all boys shouts back, and the two disappear out of view to continue their argument. Anthony leans towards me over the mobile-phone-stall counter and says confidentially, 'He's my master, you know. He gets action.' He adds, 'All of us, one love. We love each other.'

By the time the chief returns from his altercation, the swell of boys around the phone stall has grown like a rip tide. Their voices are getting louder and their conversation faster. Someone slaps somebody else and a scuffle starts. In an instant, I am completely forgotten and abandoned on my stool, like a visiting foreign potentate who has suddenly lost his hosts' interest. A voice from somewhere tells me sternly, 'You will give the chairman money to eat.'

While I am thinking how to respond to this, we hear a shriek from down the road. Some of the men run towards the noise. Anthony, until now the epitome of unassuming politeness, takes advantage of the confusion to harass one of the young women who runs the food stall behind us. As he pulls at the waist tie of her dress, she bats him away angrily and retreats to join her colleague behind a net curtain they have placed in front of their shack. The drape is all that preserves these women's dignity in an environment run arbitrarily by powerful men.

Talabi eventually emerges from the scrum of people down the road to offer me reassurance. Whenever I see his men, they will protect me, he insists. 'These are the rugged youths,' he adds. 'They can do more than you expect.'

Then he comes to my part of the bargain. I must 'satisfy' his boys by buying them a round of drinks. Even here, the coercion is gentle, respectful, mingled with protection. When one of the gang members asks for a large Guinness, Talabi roars back at him immediately, 'You will take small!' The total order comes to ten bottles of Guinness and three of Star lager. It is 10.30 in the morning. The endlessly patient mobile-phone-stall owner laughs at me. 'We will not forget you in Nigeria. All these boys around here – they are like this: "Let's go, give us something to drink".'

As the beer is doled out, the atmosphere becomes boisterous. Hands grab for bottles: even this bumper round isn't enough to quench everybody's thirst. Talabi's control of events seems increasingly tenuous, just as Asari's men behaved more wildly as my days with them wore on. Talabi leans over to me and says softly, 'Let me escort you.' He takes my hand and leads me to a waiting yellow minibus, interrupting the driver's breakfast. We drive no more than a couple of hundred yards down the road to my next appointment at an office just off Campos Square. The bill is an exorbitant 400 naira (£2); the driver has no change for a 500 naira note. 'Presidential jet,' Talabi explains with a laugh, as he turns on his heel to leave.

Back at the office, the friend who had vouched for me to Talabi says the personal connection was essential. If I had not been introduced by someone whom the boys trusted, they might have been scared of me, wondering if I was a government agent looking for reasons to arrest them. Now, they will stand by me if I have a problem, as long as I catch them at the right time and in the right mood. 'They are protective of you,' my friend says, pausing to mull over the comment, before adding, 'But then again, if you don't know it, the same boys can extort from you.'

Another friend of mine in the same office styles the boys not as viscerally violent, but as committed rationalists who have found a solution to the great oil conundrum. In this warped system where wealth springs from the ground and the till rings without fail every day, there is a much bigger return to be had in putting obstacles in people's way than in removing them. One of the boys' sources of income is spurious 'taxes' that they apply to the trucks passing through their domains, bearing consumer goods bought with the proceeds of crude. As my friend puts it, 'Everyone must have a pinch of that money. It's just like a game.'

That contest, played for the stakes of oil cash, is at its hardest and most vital in Nigeria's seething mega-city. On one level, Lagos is a macrocosm of the social forces that have pitilessly impoverished huge numbers of Nigerians and diminished their chances of survival. The World Bank estimated that in 2006 the country had a 10 per cent infant mortality rate. Nigeria ranked a lowly 154th out of 179 countries in the United Nations' annual index of human development published in 2008. The UN said Nigerians' average life expectancy was just 46.6 years, while almost a third of adults were illiterate.

For all Nigeria's substantial educated class and pockets of oil-linked wealth – the UN says the average income per person is $5 a day, compared with less than $1 a day for the average Congolese – the picture for most people is bleak. Ngozi Okonjo-Iweala, a World Bank managing director and former Nigerian finance minister, said in 2008 that more than half of the country's citizens earned less than a dollar a day. Cruellest of all is the massive evisceration of health and wealth Nigerians have suffered while the West has grown richer. The UN once estimated that average incomes in the country declined by an average of 1.5 per cent a year over the last quarter of the twentieth century.

Arriving in Lagos from Britain, where I had no children or money worries, I found myself overwhelmed at first by the degree of want and the demands made of me. I adapted, slowly and clumsily, to the lushly complex ecosystem of social and financial responsibilities. I was the manager of three office staff and a potential benefactor of many times that number of people, in my employees' families and beyond. The volume of potential supplicants multiplied even further because my business card listed my home address, which was also the *Financial Times* office. I removed the details from my card only after I received a visit from a former armed robber I'd met who asked – with, to be fair, some grace and courtesy – whether I might be in a position to help him out with a bit of money.

The quid pro quo of being confronted with this pit of need – and the uncomfortable mix of sympathy, guilt and power it evoked – was that I had never felt so in need of help and social support myself. Rarely in Britain had I had such a keen day-to-day aware-ness of my intimate and constant dependence on people outside the narrow Western universe of immediate family and friends. Confuting Margaret Thatcher's dictum that 'there is no such thing as society', social connections take the place of infrastructure in a fragmented place like Lagos. In a metropolis where there is no public water sys-tem and an air ticket can take half a day to buy, I clung to my helpers' skill and savvy as a drowning man cherishes his flotsam. It didn't take me long to realize that, in a parallel universe, they would be sitting in my comfortable seat and I in their more precarious places. In an unpredictable city where triumph and disaster are not so much impostors as daily companions, I saw both the extent of my privilege and the fragility of it.

For some Nigerians, Lagos life exemplifies how the modern state wrought by crude has become the kind of dysfunctional world depicted in Thomas Hobbes' classic book, *Leviathan*, written in the shadow of England's Civil War. Hobbes argued that, in the absence of central autocratic control, societies were doomed to exist in a state of perpetual conflict pitting all against all. In perhaps the most famous passage in *Leviathan*, Hobbes described what he saw as the inevitable and frightening results of an absence of checks on people's behaviour, when 'men live without other security, than what their own strength, and their own invention shall furnish them withall'.

> *In such condition, there is no place for Industry; because the fruit thereof is uncertain; and consequently no Culture of the Earth; no Navigation, nor use of the commodities that may be imported by Sea; no commodious Building; no Instruments of moving, and removing such things as require much force; no Knowledge of the face of the Earth; no account of Time; no Arts; no Letters; no Society; and which is worst of all, continuall feare, and danger of violent death; And the life of man, solitary, poore, nasty, brutish and short.*

It is a vision of another country at another time, but there is no mistaking the Lagosian echo. Nigeria's oil wealth gives the war of all against all an intensifying extra dimension. Living in the city is a constant reminder of the impermanence of all human-crafted structures, from buildings to systems of government. In Lagos, in whatever sphere of your life – financial, professional, sexual – you realize you must act quickly and decisively at all times, because, as a friend once put it, you don't know what will happen tomorrow.

It's a spirit that's less alien than it might first appear to the rich Western states that consume Nigerian oil and thus help finance Lagosian anarchy. As we've seen, it only takes a sharp rise in food costs or a crisis in the financial markets to remind even the most complacent Englishman in his castle uncomfortably of the possibility of sudden poverty and social unrest. The anxiety in Britain at the height of the banking crisis in late 2008 hinted at how a stability stemming from seven centuries of parliamentary tradition could disappear with surprising speed in the wake of economic catastrophe, extreme violence or environmental disaster. Perhaps a latent recognition of this underpins the popularity among Western audiences of Cormac McCarthy's *The Road*, whose post-apocalyptic ambience surpasses that even of Lagos at its most dystopian. The cities of the rich world may yet become more like Lagos before Lagos becomes more like them.

This fear of Lagos as the shape of things to come is perhaps one reason why outsiders are so anxious to dismiss it as a unique and irredeemable den of squalor and violence. The stereotype of the city's otherness is enduring, and it isn't simply a product of European ignorance, experience and prejudice. I'd regularly visit other West African countries where, having told someone I lived in Lagos, they would react with a pause pregnant with remembered fear. Then they'd tell how they'd visited it 20 years earlier and had never forgotten the awfulness of it. Ghanaians in particular, inhabitants of one of West Africa's gentler and more peaceful countries, seemed terrified of the place. They had reason to be, given that they had all been thrown out after the financial turmoil of the early 1980s.

Visitors to Lagos seem most shocked by the relentless in-your-face quality of even many straightforward daily interactions, from buying a newspaper to finding a parking spot. To call the city ruggedly opportunistic would be to grossly understate the extraordinary sense of inventiveness that sweeps across it as forcefully as the wet season rains. Oppressed for decades by dictatorship, poverty and corruption, many Lagosians have cultivated a tough and canny survivalist attitude that is an understandable and sometimes admirable response to circumstance. The crazy-looking motto on every Lagos car's number plate – 'centre of excellence' – is not as daft as it first appears.

Lagos's finest – and most bittersweet – quality is as a monument to humanity's ability to function in the most forbidding circumstances. The city, whose name means 'lakes' in the language of the Portuguese slavers who traded along the West African coast, sprawls around former swampland ill-suited to be the foundation of a modern, densely populated urban centre. After its occupation by the British in the nineteenth century, it was used as a seaport and administrative centre until independence in 1960. Its tale of rapid expansion and declining infrastructure may be common to many cities in poor countries, but the Lagos story is again amplified and clarified by the volume of oil wealth that has flowed. The country's capital until 1991, when the government fled for the newly built, spacious, manufactured city of Abuja, Lagos is expanding fast due to immigration from other parts of Nigeria and West Africa. The UN has estimated the city's population will increase from 6.5m in 1995 to 16m in 2015, making it the world's 11th-largest urban system.

The Lagos cityscape reflects the Nigerian oil narrative of ostentatious plenty amid mammoth poverty. The buildings range from apparently limitless slum districts to great estates of rich men's dwellings, including a domed basilica and a miniature version of the White House. Few of the structures, rich or poor, offer much to the aesthete. A British architect who has lived in Lagos for many years once gave me a sweeping assessment of the quality on offer. 'It's all shite,' he said. 'I know – I built some of it.'

If this seems like a patronizing slight from an unpleasant expatriate, then it is less harsh than what many Nigerians have to say about Lagos. The *Guardian*, one of the country's main daily newspapers, once carried a memorable photomontage of the city's shortcomings. The headline above the piece read: 'Lawlessness, armed robbery, gridlock, overpopulation, stench: welcome to Lagos'. Another edition of the same paper carried a full-page article on the city with the understated observation, 'These are not the best of times for Lagosians, and they know it.'

Lagos's size and rawness make it a kind of tropical Brobdingnag, casting unsparing light on *Homo sapiens'* less attractive qualities. The waste that Europeans produce in far greater quantities than Africans – but hide by burying, burning or flushing away – is left on sense-bombarding display in Nigeria's metropolis. One of the city's most distinctive and demotic perfumes is burning trash, which drifts across the lagoon waters like a grim parody of a bucolic paradise enveloped in wood smoke. In the streets, men urinate into open drains that have become sewers. In traffic jams in Victoria Island, young girl beggars tap at car windows, making terrible wailing sounds and pointing to wooden boards round their necks that explain they are deaf and dumb. Toyin Akinosho, a high-profile promoter of the city's culture, has observed that Lagos is 'becoming like London was in the eighteenth century'.

Even the beauty in Lagos often casts a shadow, reminding you constantly that you are never far from the underbelly of the human condition. Follow a splendid tropical plant to its root, and you might find a midden. Once, in a business executive's waterside office, I told him what a fine view he had. Yes, he replied, but there was one problem: corpses of drowning victims washed up now and again beneath the sill. They often remained there for some time, because the authorities failed to come and remove them.

No one in Lagos, however fat from the oil trough, could wholly insulate themselves from what people often delighted in calling the 'Nigerian factor'. Attempts by politicians and others to impose order tended to be swept away by the grand tide of humanity and events. A number of traffic lights were introduced across the city in 2004 and 2005 – complete with huge oil-company-sponsored billboards instructing drivers to 'obey traffic lights' – but some didn't work, while one switched between red and amber without ever going green. The rest were simply ignored by drivers, turning the green men into lethal false friends of the unwary pedestrian.

This Lagosian entropy was a great equalizer in an otherwise divided city. The craters on some roads were so large that they would swallow up a Nissan four-by-four and a battered old taxi equally voraciously and without discrimination, like a great whale sifting for krill. On the Third Mainland Bridge, which led to the airport, executives and commuter wage slaves alike would have to be on the lookout for cars driving on the wrong carriageway to avoid night-time traffic jams. Not for nothing did road-safety signs warn in pidgin English that 'Life no get duplicate'.

In the hands of skilful operators, Lagos's disorder and disorientation can be turned to financial advantage. The stop-start geography of the roads is well suited to the army of street sellers who underpin the city's dynamic informal economy. One day I saw a man hawking an eclectic selection of magazines that included *Vanity Fair* and *BBC Top Gear*; another held a wilted puppy up for sale in the lunchtime sun. Men even stand dangling dead rats on strings, to show the effectiveness of the rodent poison they are trying to sell. It is supply and demand economics at is most unadorned: the rich man can stop his car anywhere and wait for a smorgasbord of services to be offered him by a clamour of sweaty, competing traders.

There are important clues here to the life and philosophy of a city shaped – like the Niger Delta – by a large but unevenly suffused injection of wealth. Lagos is far from being a dirt-poor or war-destroyed metropolis like Monrovia in Liberia. Unlike a ramshackle city such as Antananarivo, Madagascar's capital, where second-hand French cars seem to go to die, Lagos seethes with expensive vehicles. Even the used cars have a certain style, with many Mercedes and BMWs sprinkled among the regulation Peugeot 405s. The high walls and razor wire in some districts of the city – including Ikoyi, where I lived – show that private élite enclaves have flourished while public life has been starved of resources.

Okey Ndibe, a Nigerian-born author and academic based in the USA who returned to Lagos in 2002 and was shocked by its degradation, describes it as a city hewn out of chaos. He doesn't mean that entirely negatively. As political and economic structures have collapsed amid oil boom and bust, so sophisticated shadow systems of organization have sprung up. Lagos is a fitting symbol of a country that only exists because outsiders interested in its resources decreed so. Forced into being less than a century ago, it is still inventing itself, often in imaginative ways. 'In a sense it's the story of Nigeria,' Ndibe says. 'Nigeria is a country waiting to be realized, Nigeria is still a fiction. That situation is seen in microcosm in Lagos.'

Lagos the oil-state megalopolis is most of all a city where neither desperation nor the dreamed-for means of escape from it are ever far from people's view. It is a place where, as a friend describes it, people run around in circles, ignoring the small things and looking for the big break, which they rarely, if ever, find. Above all, it is a chancer's paradise, rewarding spontaneity, cunning and sheer persistence.

Nowhere else in the world, after all, has someone shouted at me from the roadside, 'Hi – how are you? I want money.'

Of all the places where the multi-textured strands of this metropolis built on crude meet and tangle together, one of the most enlightening (and fun) is Kuramo Beach. Every night, music from the area's ramshackle bars diffuses hauntingly across a nearby stretch of lagoon to the upmarket Le Meridien Eko hotel. There, it drifts among the oilmen and other rich visitors who are enjoying expensive drinks and chatting up prostitutes at the poolside bar. The beach soundtrack is like a whispered invitation to come to discover the other side of a city whose islands of wealth are always being lapped at by the oceans of poverty beyond. Just like the south bank of London's River Thames during Shakespearean times, Kuramo Beach is where the city's supposedly respectable citizens can come to get their fix of vice. It is a great venue to take a vicarious sniff – and maybe even a share – of the oil riches drawn from the Niger Delta swamps a few hours' drive to the east.

Perched precariously on a peninsula of shifting sands, Kuramo Beach is an appropriate symbol of Lagos's eternal transience. It is a long line of wooden structures, stretching out eastwards on a narrow strip of land flanked on both sides by water, as if trying to join well-heeled Victoria Island to both the Delta and the impoverished vastness of the country beyond. The only amenities are those provided by the residents themselves: one described to me the makeshift hut in which people sit on benches watching Nigerian films as 'cinema central for low masses'. The posters outside advertised suggestively operatic titles such as *Oh! Had I Known*, *Cry No More* and *Dirty Image*.

The beach is the kind of seedy place that could have been a hangout for government ministers and foreign journalists in a Graham Greene novel. For a while, patrons could sit on the roof of Sammy's Bar with a beer, the resident rat scurrying round their feet as if mimicking the sea breeze caressing their heads. Then the stairs collapsed and the owner decided to turn the ground floor into a series of cubicles, each lit none too subtly with a red light.

Sex for sale is one of the things the beach is best known for. The sheer volume of it is astonishing, as the presence of an HIV/AIDS project funded by the Bill and Melinda Gates Foundation suggests. Once, as a beach regular had regaled me with sordid tales of procuring prostitutes for expats at the Meridien Hotel, he had suddenly broken into song. 'I'm a fucking P-I-M-P,' he had declared, mimicking US rapper 50 Cent. There were young women on the beach, he had confided to a corpulent friend accompanying me, who would 'fuck your belly lean'. Kuramo Beach is also the only place in Nigeria where I have been offered sex with a male prostitute. As a pimp I met at Varieties Bar gleefully put it, the beach served people on both sides of the bridge. He added, with a quintessentially Lagosian glint, that I could meet his sister too if I was interested. Everything, it seemed, could be seen, procured and enjoyed at Varieties and other venues like it – for a price.

One Friday night, Varieties' precarious wooden dance floor is shaking to the first few bars of what will turn into a revealing riff on Kuramo Beach life. There, a young man, known to everyone as Scorpion, is crouched cross-legged on the stage, like a sage. At the sound of the first beats of *lingala* – the carefree, rumba-esque music of Congo popular all over Africa – he starts to move. Dancing girls in short white skirts surround him, waggling their butts in his face; when he reaches his hand out to touch one, she aims a playful kick at him in mock annoyance.

Watched by a large, rapt crowd, Scorpion props himself up on powerful arms and starts to wiggle his own behind. He begins a breakdance-style routine, spinning himself full circle on his arms half a dozen times and concluding with a head-over-heels roll. For his finale, he folds his left leg over his right as if preparing to meditate in the lotus position. He takes a white plastic chair from the audience and does a handstand on it, punching the air as he returns to earth.

As the applause dies, Scorpion does not get up and walk away through the bar's fug of beer and marijuana. Even if he wanted to, he could not: his legs have been horribly wasted since childhood, probably by the polio that by 2008 was more common in Nigeria than in any other country in the world. He is a crippled man in a crippled city, hugely disadvantaged but very far from giving up.

When I meet Scorpion again a few days later, Varieties is bathed in late afternoon sun, as indolent by day as it is vibrant at night. I first sight him dragging himself along the sand, flip-flops on his hands and his legs slithering behind. Once under the shade of the bar's canopy, he vaults on to a plastic chair with the skill familiar from his dance routine. Seated, his legs hidden by the table, he appears a super-muscular young man, easily capable of humiliating me in an arm-wrestle or any other test of strength and dexterity.

Scorpion's face is full and his skin light, but it is his hair that attracts attention: in a country where most men shave their heads near or to the point of baldness, he wears his hair full, straight and dyed a distinctive auburn colour. The overall effect is like a cross between a fox's brush and an Elvis wig. The coiffure gives Scorpion a physical softness that mirrors his gentle speech. As he begins to talk, only the dark rings around his eyes and the sweat on his face hint at the stresses his upbeat manner conceals.

'My name is Scorpion,' he begins. 'When I was young I had a little problem.'

He was born in Congo, his father's country, in 1983. His mother was from eastern Nigeria. As a child, he felt pains in his legs and after a trip to hospital never walked again. He blames his disability on an injection he had at the hospital, but whether it was that or polio that crippled him scarcely matters now. Either way, Scorpion will be receiving no help from the state. In Lagos, many polio sufferers make money by whizzing expertly through traffic jams to beg from drivers. You always know when one of them has knocked on your door: when you look out of the window to where you expect the face of a standing person to be, you see no one there.

Scorpion recalls his mother's distress as she pleaded with him to do the one thing he could not: stand up. He told her only God could make him walk. He started to adapt to his new situation, playing 'para-soccer' with other disabled young men. Then a friend told him to do something with his future and taught him how to dance to reggae and *makossa*, another vigorous Congolese style. After several years working as a dancer in Abuja and Lagos, Scorpion would now like to study. 'Me, I would like to go to school,' he says. 'Any school, computer school. Anyone that can help me out. Finish.'

Scorpion has told his life story in less than four minutes, talking with far more economy than many people who have much less to tell. He is bright and sharp, speaking languages from at least four countries – Congo, Nigeria, England and France – with varying degrees of fluency. But he also has trouble spelling his own real name, revealing how lack of formal education has shut him out of many areas for which he is intellectually equipped.

Scorpion left secondary school in the central Nigerian city of Jos in 1998, when he started dancing. Under pressure from his mother to strike out on his own, he moved to Abuja, where he played at well-known nightclubs for about 150 naira (70 pence) a night – before Nigerian and expat audiences earning more than a hundred times that each day. This was the start of a life that has ultimately led him to local superstardom – he's even been on TV. He teamed up with a magician called Prince, who arranged for him to perform at various locations around the country, as far afield as the northern cities of Kaduna and Kano.

When Scorpion moved to Lagos he joined forces with some of the other disabled people who form one of the city's many distinctive and diverse sub-cultures. He told a new friend called Bullet that he could not work as a beggar because he was a famous dancer and musician. 'I told him that, "OK, Do you know one thing? I cannot be begging like you, understand? Because I am a superstar."' Scorpion now lives with another disabled person called Bala – a 'good, kind, man' – but thinks a superstar should have a place of his own.

Scorpion has a girlfriend, Queen, whom he met dancing. She's still at secondary school. He doesn't know how old she is, but guesses 17. She cooks for him, bathes him and would like him to stop dancing, he says. Asked if he will do this, Scorpion thinks for a long time and then begins to speak in an uncertain voice. The problem, he says, is that Queen's father is dead and her mother is not well, so she needs all the financial support she can get. She has gone this weekend to Abuja, to see if she can get some money off her brother. In Nigeria, where it is hard for a poor person to get a bank account, let alone transfer money to another city, huge amounts of people's time are spent in the physical pursuit of relatives, friends and acquaintances who are thought to have a bit of cash to spare.

It is also clear that Scorpion has another reason to want to stay on stage: his justifiable professional pride in his technical dancing prowess, his passion and his stamina. His biceps and hands, thickened by weightlifting, are huge for such a small man. He used to get round on a skateboard, but one of the wheels broke and he doesn't have the money to fix it. Now he drags himself along the street and sand on a board, using flip-flops to protect his hands. 'Somebody is a cripple, but he can dance more than the person that has two legs,' he says. 'They can never believe it. I want people to be surprised that somebody is a cripple but he will never be begging any more.' Scorpion never uses crutches.

Like many of his country-people, Scorpion would like to emigrate eventually, but before then he wants to make campaigning music aimed at a government that does not care about disabled people and others on the streets. He breaks into a segment of a song, written in pidgin English, which condemns the authorities for refusing to help beggars. It is all part of his campaign to spread his fame further, until no one can say, as Bullet once did, that Scorpion is not a celebrity. 'I need to be a superstar, everywhere, all over the country, as a gentleman,' he says, winding up to a grand oratorical flourish. 'I want me to stand as a whole, to talk to people about what they are doing, to talk to the government.'

I am struck by Scorpion's characterization of a beach society whose superficial chaos hides sophisticated shadow systems of organization and support. Just as on Lagos Island, or in a Delta militant camp, financial power flows on the beach through highly personalized structures buttressed by oil money. I wonder how many oil-worker expense accounts have drained into Kuramo Beach's informal economy. Many multinational employees seem to come to Lagos with the same attitude as the man who, on finding out that a friend of mine had moved to Nigeria with his girlfriend, had asked in puzzlement, 'But why bring a sandwich to a restaurant?'

The top dogs and paymasters in Kuramo are bar owners such as Kenneth Eze, manager of Dreams, a rival bar to Varieties down the sand. A young and ambitious man, Eze is an acute observer of the beach who understands its hierarchies and is pretty candid in talking about them. He is building a hotel down the road and he can afford to take a detached, analytical view of what goes on in the place that provides his profits. He cheerfully tells me that those who live around him are 'cursed people'.

Eze's house style, like that of the beach itself, is partly cynical toughness and partly an almost camp sense of irreverence. He has the engaging habit, not uncommon in Nigerian English, of referring to people of both sexes as 'my dear'. He sometimes uses the phrase to evoke a sense of shock. 'When you are living here, my dear, it's hell,' he confides to me. 'Because there is nothing. Would you want your daughter to be a prostitute?'

Eze grew up at the nearby Kuramo Beach police barracks, the son of a police transport officer father and a mother who traded frozen food. He dropped out of university. When asked why, he replies simply, 'Stubborn.' Asked what he means, he roars with laughter and declines to elaborate. 'I grew up on the ground,' he says. 'There is nothing I don't know about this neighbourhood – nothing.'

Like the area-boy leader Talabi, Eze is head of the hustlers in an environment well populated with them. He shrugs his shoulders at the behaviour of the lively young men in the car park, who direct guests to parking spaces and bars in exchange for a pay-off. Better the small-time villain you know, he suggests, than the big-time criminal you don't. 'My dear, you cannot steal here,' he says. 'It's like a village. There is one tradition I have come to understand about robbers: they will not steal where people know them.'

The low-level spivs collect money mainly from the prostitutes who work the beach, he says. Some of them also sell drugs. He does not do so himself, although he had trouble once after someone planted cocaine on his premises and called the police. He survived, just as the real dealers always will. For drugs are not a poor man's recreation: some of the biggest users are the people who promulgate the laws banning them. A prominent person will come to the beach and share his weed with the hustlers. 'Something about cannabis,' Eze muses. 'Everyone who smokes it feels they are one blood.'

But, just like a drug hit, this fellow-feeling is impermanent and has tightly defined limits. People are not naturally united, Eze says, but only when there is a common difficulty to be overcome. For Eze, as for Scorpion and others, Kuramo Beach is a wobbly bridge to better things. Eze plans to get married next year. He laughs when asked if his wedding will take place on Kuramo Beach, replying with a hint of steel that he will stay well away. If he came, he says, the car park boys would be bound to 'intrude'.

The Kuramo Beach that Eze evokes has a harder centre than the accepting environment described by Scorpion. What Eze recounts is more like what a Ghanaian acquaintance – in another reminder of Hobbes – calls 'crab in a basket' syndrome, with each person trampling on others to get out. Eze himself is an operator who hides an entrepreneur's alertness beneath booming bonhomie. He is coy when asked how much money he is making from Dreams. 'I won't tell you that,' he says. 'That's what makes me a businessman.'

When I suggest that getting him drunk might loosen his tongue, he laughs incredulously. His reply is in keeping with the larger-than-life quality of all events on Kuramo Beach. 'Get me drunk?' he roars. 'Is that possible? My blood system is alcohol. I drink and drink and drink to the next morning.'

The deeper I looked at the apparent random complexity of Kuramo Beach and the other enclaves of this city built on oil wealth, the more I found a revealing simplicity at their heart. If crude is king in Nigeria, the cash it generates is the monarch's largesse, its limited and careful distribution ensuring that social unrest is checked and the status quo just preserved. In their focus on the possible, rather than the ethical, Talabi, Eze and the rest of the army of freelance Lagos entrepreneurs would not have looked out of place in a City of London trading house or hedge fund. They had moulded little islands of prosperity, but they were being battered by the sea of insecurity around them. In Lagos – as elsewhere in the world – the economic wealth never trickled down to those at the bottom in anything like the quantities a neo-liberal economist might have predicted or hoped.

As the oil price climbed steeply in 2007 and 2008, the new Lagos State administration of Governor Babatunde Fashola tried to bring a degree of order to its sprawling urban charge. Highways were paved, area boys quietened and the rules of the road applied a bit more rigorously by special squads charged with policing one-way streets. A main junction near the airport acquired a traffic light system complete with time counter, while trees in upmarket Ikoyi were adorned with signs detailing their species, their common names and their uses. Whether Lagosians viewed these investments as civilising essentials or elite frippery probably depended on where they sat in the city's enduringly stark economic caste system.

Oil money was still sharpening the colours of Lagos's social contrasts, illuminating fault lines that stretched far beyond Nigeria's shores. Just as the difference between Nigeria and war-razed Liberia was one of degree, I felt the same could be said of Lagos and London. Gazing out from the shifting Kuramo sands at the lights of the ships lined up to enter port, my homeland suddenly seemed closer at hand than I might have anticipated. Eze found the right words to explain why I found this captivating beach outpost in an oil war so familiar. Life there was, he said, simply a 'money-go-round'.

5

FUEL THE BIKE, FUEL THE RIDER

In a country pillaged for years by dictators, styling yourself as a senior military officer might not seem the smartest way to win popularity. But at Charly Boy Oputa's Lagos compound, posters proclaim him 'general of the people's army'. The personalized number plate of his Nissan Pajero four-wheel-drive bears the same message, while a kitsch photomontage outside the house shows him on a motorbike leading hundreds of other riders as if into battle. The illustration carries another one of his nicknames: 'area fada' – area father – placing him in a pseudo-paternal role above the area boys in the hierarchy of the metropolis.

The son of a senior judge, Oputa is variously a music promoter, motorcycle enthusiast and spokesman for the voiceless. Perhaps most famously, he campaigns for the interests of the drivers of the motor-bike taxis, known as okadas, that allow Lagos to function during its epic gridlocks. Whole societies exist around the bikes and buses that move millions of citizens through the metropolis's warren of bridges, high-rises and slums. Lagos's roads are one of the main places in which all the social classes under the rule of oil meet and skirmish. When I lived in Nigeria, if you arrived at Lagos airport and drove into the city, one of the first sights you would see would be a giant billboard showing Oputa astride his motorbike, as if in preparation for a roadway joust.

Lagos's transport wars are a telling illustration of how free-flowing resource wealth can add to the growing pains of a country trying to make the transition from autocracy to democracy. Just as Nigeria's creation coincided momentously with Churchill's switch of Britain's battleships to oil, so the country's return to civilian rule was accompanied by an epoch-changing recovery in the world oil market. The international price began a long climb from $10.75 a barrel in February 1999, three months before the end of dictatorship, to more than $130 a barrel in 2008. Suddenly, the Nigerian politicians and officials who ran the weak new civilian institutions had access to bags of cash for which they would no longer be held accountable by the military leaders who had ruled for the previous 15 years. The Lagos transport system – never exactly clean to start with – became a classic cascade of pay-offs, as people who suffered from the corruption of those above meted out similar treatment to those below. Once again, Nigeria was a stark exemplar of the sometimes grotesque power relationships that define our unequal world.

My expert guide through the Lagos transport battlefield is Sina Odugbemi, a lieutenant of Oputa. Odugbemi is tall, slim and has a face for fun, its features seemingly poised perpetually between querulousness and laughter. I set off with him and three okada men for the short journey to the nearby district of Somolu, where we are to meet members of a drivers' union. Rugged, poor Somolu ought to be prime okada territory, but our casualties in reaching it are high. The first driver drops out complaining about the rough roads, the second runs out of petrol, while the third is chastened after falling into a large pothole. He limps home, like the sole survivor of a humiliated army unit. Odugbemi, shaking his head in amused admonishment, is not impressed.

It is only just after 7am, but Somolu is already well into its day. Life starts early and alertly in this part of the city, as in many others. Street vendors sizzle dough balls to fuel a public whose smart dress belies the filth and decay around them. Most people have about them an apparent robustness that their surroundings lack. The commuters stream past a Volkswagen Beetle that lies wheel-less by the roadside, collapsed and helpless as a new-born foal.

Opposite the area's Al-Barka mosque, a lean-to houses the headquarters of the Palm Grove Unit of the Motorcycle Operators

Association of Lagos State. The palms have long gone, if they were ever there in the first place: the only reminder of wilderness here now is a chained monkey belonging to one of the okada riders. The animal paces restlessly through its small slice of urban jungle, its long tail sweeping perilously close to the fetid waters of an open drain that runs parallel to the road.

We are greeted by Olufemi Orimogunje, chairman of the Palm Grove Unit. Orimogunje is raffish, if a little shop-worn by age. His short hair is clumpy, his face pitted, and his long, thin moustache would need only a small tweak to make it resemble an okada's handlebars. A trained biochemist who started riding bikes in 1997 after he couldn't find work as a teacher, he is part of the mélange of unemployed professionals, illiterate youths and natural-born chancers who find their way on to the okada circuit. 'Because of the economic condition of the country, I wasn't able to manage that teaching profession,' he says. 'So I entered the okada line.'

The okada has risen relentlessly as other industries have withered in the face of the obsession with crude that has both squashed Nigerians' job opportunities and forced them to create their own versions of the industries, services and infrastructure Westerners take for granted. As unemployment has increased and businesses have failed, many educated, trained people have turned to okada-riding, much as middle-aged men sloughed off by Western capitalism become security guards and minicab drivers. The sheer number of okada riders who swarm deafeningly round Nigeria's urban centres is a sign of both a broken transport system and the difficulty of finding jobs elsewhere. As Orimogunje puts it, 'A tailor who doesn't have any cloth to sew with will become okada rider. A mechanic who doesn't have any tools will become okada rider. Unemployed graduates will become okada riders.'

It doesn't take a visitor to Lagos long to see that okadas provide the essential grease that stops the coarse cogs of the city's social life damaging each other beyond repair. Like many Lagosians, I found okada riders brave and infuriating, capable of great canniness or astonishing foolhardiness – or, quite often, both at the same time. If you are in a car or bus, you will curse their incessant horn-blowing and gadfly dodging across lanes; if you are late for a meeting, you will silently cheer your okada driver as he rides roughshod over traffic laws to get you to your appointment in time.

Orimogunje admits immediately that many riders routinely break the law. They fail to follow rules that say they should wear helmets, carry only one passenger at a time and be properly certified and insured. Orimogunje justifies this on the grounds that drivers can't afford to do these things in an industry in which profit margins can be as small as a pound a day.

In their wilful flouting of the rules, okadas are simultaneously responding to and helping to perpetuate an economic system in which a lack of regulation has allowed debatable charges, hidden taxes and extortion to flourish. Each rider pays 20 naira (about ten pence) to the three local governments where the union's members operate, Orimogunje says. A further, negotiable, sum goes to the police to forestall both legitimate arrests and vexatious harassment. Paying off the authorities also helps deter area boys from extracting cash.

In other words, the police take bribes from okada riders, who are in turn breaking the law to do a job that probably keeps them away from more serious crimes. It is one of many reminders that in Lagos, the spectrum of behaviour covering law enforcement officials, ordinary transport workers and criminals is often narrow indeed. All feed from the same pot of cash generated by people's need to get around. If okadas were banned, or at least more tightly regulated, it would be disastrous, says Orimogunje: apart from the logistical problems it would cause, it would mean 'sleepless nights for everyone' because sacked riders would turn to robbery. 'The crime rate would increase,' he says, 'because okada riders save many souls in employment.'

This most individualistic and accidental of professions has generated a paradoxical sense of solidarity that cuts across ethnic and religious lines. While it would be foolish to romanticize this – team spirit is, after all, as the footballer Steve Archibald once put it, an illusion you glimpse when you win – many people have noticed how riders will stick up for each other, often at some danger to themselves. In part this is a response to the large number of accidents involving okada riders: the secretary of Orimogunje's branch was killed on the motorway the previous year. 'When we are riding an okada, we are one,' he says. 'We protect the interests of the okada rider – whether wrong or right.'

In one of the most eye-catching examples of group action, okada riders rioted in the Ikeja district in 2005 after one of their number crashed into a car belonging to a military officer, who promptly stepped out of the vehicle and shot the rider dead. After the murder, okada riders destroyed properties worth almost 4m naira in solidarity, Orimogunje says, with some pride. He adds, 'They are even threatening to ban okadas in Ikeja because of that.'

Orimogunje styles this commonality of feeling not as a realization of shared interest and the power of the collective, but as a visceral process born of a deep anger at the dangers of okada riding and the wider injustices pushing people into doing it. 'It's a system you enter by force – force of the economy,' he says. 'That's what always brings our solidarity – people united by frustration. It's like particles attracted by a magnet.'

Where such feelings fail to carry along all the riders, the union has to step in as enforcer, he says. If people don't pay their union fees, their okadas are taken from them. If they still don't pay, the riders are taken to the nearest police station to face the 'wrath of law'. In other words, the unions seek the help of the very officers they despise.

As Orimogunje talks, I notice a lidded metal cup under his chair. Is it tea? I ask. There is laughter. It's not tea, he says, but *paraga* – an eye-watering mixture of spirits and herbs. He takes his first slug of *paraga* at 7.30am, before he has any food. Okada drivers are renowned consumers of cigarettes, Indian hemp and *ogogoro*, a powerful home-brewed gin made from the juice of raffia palm trees. Odugbemi notes wryly, 'Fuel the bike, fuel the rider.'

Orimogunje describes how the sprawling okada industry has become big business. Wealthy men buy bikes for a few hundred dollars and then sell them to riders at steep hire-purchase rates. It turns out Orimogunje himself is rather straddling the divide between shop floor and management. He doesn't ride any more, but he owns three okadas and takes 500 naira a day from each of the riders. He claims he is more humane than other owners, because he pays for repairs, whereas they do not.

Orimogunje is less forthcoming about how much supplementary income he enjoys from his union activities. At first he claims the union hasn't started remunerating its staff, even though it was set up more than ten years ago. He then says the branch's income is only about 2,000 naira a day, although he still won't reveal what cut he takes. I persist in asking about his extra earnings, but he continues to avoid the question until the end of the interview. Then the *paraga* is passed around and some commemorative photos are taken. Okadas are here to stay, says Orimogunje; they are part of the system and Lagosians are 'loving it'. 'The role of okada riders is to make life far more social for people in Lagos State,' he says.

After we leave, Odugbemi observes how reticent Orimogunje was about how much money he earned from the union and what he did with it. Even as Orimogunje complained about the corruption of wider society, he himself was running an unaccountable organization. Nigeria is like that 'from top to bottom', Odugbemi says ruefully, even after the return of civilian rule. The power structures of the okada business are a 'mirror image' of how the rest of the country is run, for all the periodic attempts by the authorities to tighten the rules, such as by forcing riders to wear helmets.

Back at Charly Boy Oputa's place, it doesn't take long for some of the tensions deplored by Odugbemi to break the surface spectacularly. After a late breakfast of rice and beans from a nearby stall, Odugbemi sets me up to chat with another group of okada riders, including another union official. We go to a small outbuilding, from where we can hear the rumble of the main road. Inside, the three okada men and I sprawl on the fake leopard-skin chairs and chat.

The union official, Leonard Agumba, speaks first. Agumba, his plumpness not quite concealed even by a generously flowing shirt, is as coy as Orimogunje about saying what his income is. He won't even say how many members there are in his Gbagada union branch, which he chairs. Eventually he admits that about 200 naira is left out of each day's takings, half of which goes to him as commission and the other half to the union's coffers. He says people who don't pay their subs are charged a 100- or 200-naira fine. The money is used for social welfare: they gave 400 naira to a boy who lost both legs, and 2,000 naira to a man whose wife had died.

I asked what happens to members who don't pay up. Agumba's sidekick, Innocent Adio, answers, trying to be helpful – and earning his name. 'They will force him and beat him and humiliate him,' he says. The chairman anxiously corrects him, 'Not beat him. They will seize his machine.'

The third rider in the room, Timothy Dada, cannot contain his impatience at these claims. Agumba and his allies are not part of a disciplined organization, he says, but a group of area boys calling themselves a union. They are self-appointed extortionists. 'If your wife dies, they will not give you anything,' he says. 'No welfare – it's just a group of people working for themselves.'

Agumba reacts angrily. 'Why are you accusing me, my friend?' he says. 'Stop it now.' The union is properly registered, he says, and operates with the state government's full consent. His sidekick Innocent intervenes again, once more enraging his boss. Okada riders are afraid to fight the union bosses, he says, because they are too powerful. 'The best thing is to submit,' he says.

It is too much for Agumba, who announces that he and Innocent will be leaving. As they reach the threshold, Dada starts to laugh derisively. Agumba turns back inside, shuts the door and harangues Dada. 'What is wrong with this man?' he scolds. 'Are you mad? Someone calls you for interview and you talk rubbish.'

The conversation has become a surreal study of how – in a country that is 'too big, too wild and too many' – even the powerful can never quite silence dissent and irreverence. Innocent, Agumba's unintentionally provocative assistant, chips in yet again, saying that people are as afraid to question union management as a servant would be fearful of challenging his master. Agumba, who is looking less masterful by the minute, makes a second attempt to get out of the door, this time successfully, leaving Dada the chuckling winner of this little skirmish. 'These okada riders – they know me very well,' Dada says. 'They are afraid to tell the truth.'

Dada now takes the floor to outline his own circuitous journey to the okada trade. Dressed in jeans, sweatshirt and a baseball cap, he is a good-looking man, well-toned from daily trips to the gym. He is Nigerian but grew up mostly in neighbouring Cameroon. He has pursued trades as diverse as photography, topographic mapping and repairing electrical transformers. After working for a while in Cameroon, he saved some money and decided to try life in Europe. He travelled to Germany but was turned back at immigration because, he was told, his papers were not in order. He was not even in the country for 24 hours, he reflects ruefully.

His next stop was Nigeria, although he found it hard to get work there. He set up as a freelance photographer – the kind often seen at weddings and social events, taking pictures of guests and trying to sell them afterwards. But there was only really work at weekends, so he started driving an okada as a way to earn money during the week, to support his wife and three children. Like so many of his fellow Nigerians, he has discovered that a single job and single trade are not enough to survive.

For all his anger with the union management, Dada shares with them a strong sense that he is an okada man through duress rather than choice. Both top officials and ordinary members swim in the same wide and deep pool of discontent. Dada says that economic conditions are worse now than they were during the decades of dictatorship. Of all the lessons Nigerians have had to learn since the oil started flowing, perhaps the hardest and most important is that government will give them almost nothing in terms of services. As Dada puts it crisply, 'No road, no water, no light – no difference.'

Once Dada leaves, it's time for my second outing of the day courtesy of Odugbemi. I had wanted to spend the afternoon following one of Lagos's classic bus commuter routes, catching the evening rush hour and seeing just what pitfalls lay in the way of a mass transit operator in Africa's largest city. Odugbemi had put me in touch with a driver, whom he said I should meet at a stop just across the road from Charly Boy's place.

Making the rendezvous is by no means as straightforward as it sounds. The road is a carriageway several lanes wide, densely populated on both sides, but there are no pavements, no traffic lights, no bridges and no subway. People scurry across to the central reservation, hoping that waiting for the car in the slow lane to pass hasn't put them on a collision course with the initially more distant but faster cars on the inside. It is a test of geometry, relativity and ingenuity. For all the destruction wrought in African countries by famine, war and disease, moments like this serve as a reminder that apparently mundane daily risks are among the most prolific killers.

Waiting on the other side is a big yellow beast of a bus known as a molue. These are the alpha males in Lagos's menagerie of public transport, dwarfing the okadas, taxis and yellow minibuses known as danfos. Many molues are made by the Tata Group of India, where they provide a ponderous but reliable method of crossing vast stretches of country. They are terrifying if you are unfortunate enough to be passed by one while you are riding a scooter, bicycle or rickshaw. It is hard to know whether to be reassured or disturbed by the admonishment painted on to the back of our bus: 'Safety is of the Lord.'

Inside, this leviathan of the road is decorated in suitably hyperreal fashion. Its dashboard is draped with a bank of luxuriant woven rugs, laid out like animal fleeces awaiting the attention of a Jason or Hercules. A cross and a sprig of leaves hang from the giant rear-view mirror, which is sticky-taped in place at both ends. The driver looks strangely shrunken and powerless compared with the outsize objects around him: each time he turns or changes lanes, he wrestles with the wheel as if roping a steer.

The rest of the bus is crammed, with three people hanging from both front and back doors. The floor is strewn with orange pips and rind: evidence of the hawkers who find a way in, no matter how bad the crush. One man is trying to sell a restorative blood tonic for 200 naira. 'I love my wife. I love my children,' he declaims, trying to pull potential customers into the slipstream of his lust for life.

Perched a couple of rows behind the driver is Chiguzor, my contact from Odugbemi. Chiguzor, a gently-spoken man, is a driver who says he is not able to work because he needs to keep a low profile. He is in trouble with the National Union of Road Transport Workers – the organisation the area boy chief Talabi works for – for complaining publicly about the way it extorts money from bus drivers. This has created problems for his family that will require more than a blood tonic to remedy: union officials have beaten him up twice and told him that they know where his wife and children are. 'The national union want to kill me,' he says matter-of-factly. 'As I am talking to you, my children can't go to school, because I can't drive.'

Earlier that day, seated on the same fake leopard-skin chairs that the warring okada riders had occupied, Chiguzor had taken Odugbemi and me through the economics of running a molue in Lagos. The buses are managed on a kind of franchise basis. The drivers and conductors collect the fares for the day, cover all the expenses, give the vehicles' owners a cut and then keep whatever remains for themselves. So any problems on the road, including extortion, hit the bus staff straight in their pockets.

Chiguzor reckons his molue brings in revenue of about 32,000 or 33,000 naira (£150) a day from a route that takes it from near the international airport to Obalende on the edge of the rich offshore islands. It is a route heavily used by commuters who live on the mainland but work in the business districts. About 8,300 naira of the takings go straight to the owner as a flat fee. Another 12,000 go on fuel, the price of which more than doubled in two years as the government gradually removed a longstanding subsidy. Another 8,000 or so will go to paying off National Union of Road Transport Workers officials who police the route and board buses at will. That leaves the small sum of 4,000 naira.

At this point, Chiguzor pauses in his account before adding, with perfect comic timing, 'Then the police will collect 2,000 from that.' I can't stop myself laughing at the dark absurdity of it all. I apologize, but Chiguzor also laughs and says, 'It is funny.' By now, everyone is laughing, including Odugbemi, who cries out, 'Painfully funny.' Nigeria's situation has turned many of its people into connoisseurs of the ridiculous, who have made self-deprecation a national pastime.

Once the laughter has subsided, the disturbing reality remains. According to Chiguzor, the busmen will be left with about 2,000 naira

– well under £10 – of which half will go to the driver and a quarter to each of the conductors. It's a meagre return for a hard day's work.

The cash cascade shows how the bus industry has become big business to all those who feed off it. The molues may rule the road physically, but the uniformed and un-uniformed touts who hop on and off to take money exercise the real power. Chiguzor believes that many of them have the backing, implicitly or explicitly, of state government officials. In return for helping rig elections, their reward is to be allowed to rule the streets. The road transport workers' union used to take 17,000 naira a day, but agreed to halve the amount after Charly Boy and Odugbemi intervened on the bus drivers' behalf. The extortion may be arbitrary, but it can also be arbitrated.

Chiguzor says that if the bus drivers ask police to arrest the touts, they may oblige, before releasing their catches in short order. On one occasion, when some drivers he knew complained to the police, they were themselves locked up and had to be bought out at a cost of tens of thousands of naira. As an ordinary Nigerian citizen, the last thing you want to do in a dispute is to involve the police, particularly if the quarrel has potential political repercussions.

Once our molue is on the road, the many-coloured uniforms of the various transport authorities give a kaleidoscopic aspect to the daily struggle for power and money. At Osodi, we see the first maroon-trousered employee of the Lagos State Traffic Management Authority – known by Chiguzor as 'MOT' – wiping sweat from his forehead. Nearby is a National Union of Road Transport Workers official, dressed in green and white and armed with a cane. Next, three female police officers get on the bus, two in black uniforms and one in plain clothes. Chiguzor playfully bats away the hand offered to him in greeting by the plain-clothes officer. He then has a mock fight with the taller of the uniformed women, after which the three officers laugh and leave. Chiguzor says they asked for 500 naira. 'We did not give them – because we know them,' he says.

Asked how he manages to deal civilly with people who are trying to take money from him, Chiguzor's answer is considered and realistic in a way that outsiders in African countries all too often mistake for childlike fatalism. 'If you continue hating them, that makes it worse,' he says. 'They will just drag you to the station immediately. They will just say, "That guy is rude." And whatever they say, their boss will believe. That's why we try to make friends – so sometimes, if we plead on the road, they will allow us to go.'

Luckily for me and my fellow travellers, our driver is skilled in the evasive actions needed to avoid unwanted official and unofficial attention. Frequently he slows at bus stops but never quite halts, leaving passengers to judge the best moment to jump off and on. It is as if the vehicle is following the quantum mechanical law that predicts sub-atomic particles can never be completely still, because that would violate Heisenberg's uncertainty principle. The Lagos bus version of the theory is equally persuasive: the shorter the stopping time, the less opportunity non-fare-paying visitors have to climb aboard.

Chiguzor says he and other drivers are finding it 'very, very difficult' on the roads, because of the gangsterism that has flourished since the end of dictatorship in 1999. Nowadays, politicians need to use their access to oil money to pay police and thugs to rig elections and keep them happy until they are needed for the next polls. In other words, civilian rule means more mouths to feed and more extortion. Democracy is no panacea for most people's everyday struggles.

Approaching the Osodi bridge, the bus slows to a crawl to take on more passengers. It can't pick people up on the bridge because Lagos State Traffic Management Officials are waiting to pounce. By the time we have juddered up the hill, we have about two dozen extra fares. It is a virtuoso piece of driving.

The bus performs a similar trick next to the Third Mainland Bridge, meandering in the slow lane as cars whip past. Around us, Lagos looks beautiful: the sails of the fishing boats on the lagoon are billowing, while canoes are clustered around fishing traps. A lone goat grazes on a litter-strewn patch of grass, in a tiny square formed by four major roads that converge at a spaghetti junction. Like most of the city's human population, it seems the animal is making the best of a difficult day.

At the Obalende terminus, the bus staff have to pay off an official of the road transport workers' union, who is waiting at the bus stop. He explains that he takes 40 naira for every non-union bus he sees, because the union runs its own buses in competition. 'All fingers are not equal,' he shrugs in explanation. He has nothing to say about Chiguzor's claims of extortion, referring questions about them to the union's head office.

By the time we start off on the next circuit, the rush hour has begun. The harshest of the sun is over, but the molue is now permeated by the day's sweat. The conductor runs short of change, so he starts giving people cards with numbers on to show he owes them money. The driving, too, has acquired a less controlled, more urgent edge. On the flyover that looks down on the dense commerce of Lagos Island, another molue refuses to let us into his lane. Our driver shouts at his adversary, 'If you won't let me come by, I will finish your motor!'

In a traffic jam by the shoreline, street vendors start to filter between the cars to take advantage of their temporary immobility. A hawker drops some glass tumblers, which smash on the ground. As he's salvaging what he can, our driver turns the steering wheel to avoid him. In doing so, we almost crush two vendors standing next to the rail on the other side of road. One bangs the side of the bus in alarm and frustration. No one is really at fault, but Lagos's crowding and wafer-thin margins of error mean a small accident has almost turned into disaster, like the apocryphal butterfly-wing-beats that ultimately cause the hurricane. What's most interesting is Chiguzor's reaction, which is to turn this microcosmic street incident into a political comment on the state of the nation. 'We have bad government in this country,' he says. 'We have money. We have everything. But we have bad government.'

As we continue to skirt the heart of Lagos Island, a sweaty young man in street clothes gets on and sits behind the driver. After a short time, he takes 100 naira from the conductor. He won't talk about what he's doing, other than to nod briefly in acknowledgement when I ask if he's collecting money for the union. He walks to the back of the bus and exits, appearing moments later at the front door, where he calls out, thrusts his 100 naira towards me and then pulls it away again. Chiguzor tells me he was saying that, although he knows people do not collect money like this in Britain, it is the habit in

Nigeria. It seems a defensive gesture, betraying an unexpected sense of embarrassment, even guilt.

The next visitor is a young evangelical Christian, who begins to preach as if in instant spiritual reproach of the bribe-seekers. Pentecostal church leaflets poke out from a burgundy Bible that he holds in front of his face as he proselytizes. It's like a masquerade, the faceless word of God booming from the front of the bus. 'No matter what you want to have today, no matter what disappointments you have, I want to tell you tomorrow is another day,' he shouts. 'The Lord has a programme for you today. The Lord loves you.' He starts to sing, prompting a sizeable minority of passengers to join in, whether out of faith or tedium. Chiguzor is one of them. Eventually the preacher stops and continues with his sermon, which he ends with a warning, 'I want to tell you that we are entering a dangerous mood of the year.' He says a few amens, shakes hands with the passenger behind me and alights. His exhortations to moral living have fallen on at least one pair of deaf ears. As he finishes speaking, a man in a smock and red fez-like hat gets on and takes 20 naira from the conductor.

The next visitor is a stoned-looking man in a grubby white T-shirt. He leans unsteadily as he sticks a fistful of 200 naira notes in the driver's face. Chiguzor approaches him with a kind of cautious respect, shaking hands and asking, 'How's work?' Eventually the conductor comes up and asks a more relevant question: 'How much?' He gives the 200 naira the man demands.

In the busy district of Orile, a uniformed police officer gets on and talks to the driver. He then gets off again and loiters by the door. Suddenly he looks up, sees me looking at him and ducks away from view. The conductor gets out, calls him back and gives him some money out of my sight. As the bribe is paid, a police car with a siren passes us on the other side and forces us to the edge of the road. The car bears the force's new motto: 'To serve and protect with integrity'.

A hawker climbs aboard to sell glucose, even though two things this bus does not lack are energy and stimulation. The owner of the bus also arrives to collect his cut of the money. He is resigned on the question of extortion, saying that if those who run the bus did not pay they would have problems. As he counts the day's takings, a uniformed Road Transport Workers' Union official gets on, shouts out a demand for money and gets off again. The owner shouts back: 'What's the problem?' He alights, tapping his head with his finger to indicate that he thinks the guy is crazy.

Chiguzor doesn't see any easy way to stop the endless cycle of pay-offs. He certainly doesn't think a popular uprising will end it. Weary passengers want an easy life, so they often tell bus staff to give touts money and avoid more serious trouble. 'That's what's killing us in this country,' Chiguzor says. 'No one will come out and say, "what these unions are doing is wrong."'

I reflect on the truly staggering quantity and variety of corruption that Chiguzor and his colleagues have to deal with every day just to get by. Almost no one with powers over transport was to be trusted not to abuse them, it seemed. The point seemed tacitly acknowledged in the post-2007 Lagos state government's efforts to bring a dose more order to the unruly system. The administration provided a fleet of new public buses and regulated okadas more tightly, leading to an unprecedented outbreak of helmet-wearing among drivers.

What will be slower to change are the daily pay-offs on the roads and the blanket suspicion many Nigerians feel for authorities ranging from lowly transport officials to the governments that have run the country during its short life. The bus and okada bribes I'd seen – rising all the time with inflation and fare increases – were but grains of sand compared with the state graft that Nigeria has suffered since the world struck oil there. That corruption has – with the complicity of Western countries and companies – helped drain the nation of the $350bn or more estimated to have been wasted through fraud and mismanagement since independence. Lagos's streets were at the bottom of an inverted pyramid of graft whose vast top, I realized over time, stretched across the world. Nigeria hardly had a monopoly on financial wrongdoing, but the range and magnitude of oil-fuelled misdeeds had once again made the country a telling example of a universal problem."

At last, the molue reaches Osodi and turns again for Obalende. The traffic is much calmer in this direction, as we are travelling into Lagos's business districts rather than out of them. The driver produces a handkerchief and has a richly deserved wipe of his face. Among the passengers, one young woman reads a romantic novel, while a man studies a copy of a book called *Search the Scriptures*. Some people look stressed, while others doze. We swerve to avoid a danfo whose tyre is being fixed in the slow lane, causing the truck on our other side to blare at us in protest. My bus journey is nearing its end, but, for the leviathan of the road and its stewards, the exhausting daily combat with cars and corruption will go on and on.

6

THE DISCERNING GENTLEMEN

Even in the thick of a world property market crisis, a flat in the Water Gardens, west London, is worth a bob or two. The high-rise apartment block is a haven from the mercantile cacophony of the adjacent Edgware Road. The development takes its name from its fringing landscaped pools decorated with fountains, aquatic plants and jumbo koi carp. The tableau is described by one property website as 'almost unique in London for quality of design in a confined area'.

On the top floor, at the end of a small hallway decorated with a David Hockney print, is the triple-deadlocked door to flat 247. Inside, the stairs are marbled and the balcony adjoining the vast living room offers views of Hyde Park and the BT Tower. In the master bedroom, the wooden frame of a four-poster bed still stands in the centre of a floor that is a bomb site of clothes and other personal belongings, including a pamphlet entitled 'Energy, Environment and Disaster: the Niger Delta Experience'. 'This,' says my guide to the now-vacated flat, gesturing towards the mess, 'is the room in which a million pounds in cash was found.'

That money – £915,000 in sterling, euros and dollars, spread all over the room and even stuffed in the pockets of a dressing gown – was discovered by Scotland Yard investigators in September 2005 as part of one of the ghastlier and better documented of the suspected corruption cases spawned by Nigeria's oil boom. The owner of the gown – and the Water Gardens flat – was Diepreye SP Alamieyeseigha, then governor of Bayelsa State, where so many of Nigeria's battles for oil wealth have taken place. British and Nigerian police claim he took millions of pounds in stolen money and bribes during his time in office and salted it away in Britain, in both property and accounts at many of London's biggest banks. In 2007, the High Court in London ordered Alamieyeseigha to give up many of his British assets, including the Water Gardens flat, to the Nigerian government after he received a two-year jail sentence in Nigeria for making false asset declarations and allowing money-laundering by his companies. Once again, the threads of Nigeria's venal oil industry had entwined with the institutions of the British Establishment, showing that the Delta's apparently distant troubles were not so far away after all.

The Water Gardens flat is a case study in how Nigerian oil state officials have profited from finding safe havens for their wealth in leading Western cities. On a wall in the living room, across from an empty Lancôme perfume box, a photo of Alamieyeseigha is displayed alongside a metal engraving of Jesus addressing the Last Supper. One of the governor's trademark black Panama hats sits on a table opposite. It's from Edward Bates of Jermyn Street, which, according to its corporate website, supplies 'stylish hats and caps to discerning gentlemen'. Even the toilet is a cut above the norm: it's the first I've seen in a private home that has a urinal as well as a bowl.

The flat's collection of books and papers testifies to a lifestyle full of chutzpah and internal contradictions. Many are religious works, with titles such as *How to Trouble your Trouble*, *Smite the Enemy* and *Satan Get Lost!* Others – including a brochure for luxury boats, with a picture of a speeding craft and the catchline: 'follow no one' – suggest a mind not immune to material pleasures. In the hallway sits a document shredder, its bin a tantalizing quarter full.

Among the detritus is a copy of a book written by the governor, entitled *Managing Youth Unrest in Nigeria: a Holistic Approach*. It is dedicated to 'the youths of Nigeria, east or west, north or south, and to all those who stood by me in the dark days of youth unrest in Bayelsa'. Its 201 pages are turgid, but near the end is an intriguing section giving the governor's thoughts on developing a 'culture of probity and accountability'. The conclusion could be his epitaph:

To the average Nigerian it does not matter what a person does to become rich. Once someone has acquired wealth, they are practically worshipped. This attitude encourages people to do anything to get rich quick. Another is the habit of cheating and cutting corners – so accept-able it becomes institutionalised.

My first encounter with the colourful life of DSP Alamieyeseigha had come more than five years previously, on the day before his April 2003 election to a second term in office. I was visiting Bayelsa to do a story for my newspaper, after reports of trouble there during voting in parliamentary elections the previous week. As I passed billboard after billboard showing large and intimidating images of the thick-set, full-faced governor in his black hat, I felt I was travelling to a Mafia anointing rather than a democratic poll. Alamieyeseigha was openly using official property to campaign, suggesting he had few qualms about blurring the boundaries between his personal affairs and those of the state. On the gates of Government House in Yenagoa, the state capital, a banner proclaimed: 'DSP: your knowledge is our strength.'

I drove up the road to the town of Sagbama, where burnt-out cars surrounded one of the governor's lodges. It wasn't quite in keeping with the state's self-declared status as the 'pride of the nation' and the 'glory of all lands'. The main building, still festooned in Christmas decorations, had also been gutted by fire. Officials told me this was the product of a little local difficulty between ruling party supporters and their opponents. The parliamentary elections had resulted in eye-openingly large majorities for the governor's People's Democratic Party. In the House of Representatives seat of Brass/Nembe – site of the 1895 palm oil uprising – the official results showed that the 131,335 valid ballots cast for the PDP had exceeded the 129,535 registered electors.

I found evidence to support opposition supporters' claims that the Sagbama lodge was being used as a headquarters to rig the polls. On the ground outside the building, I saw an official document used for tallying spoiled and rejected ballot papers. A saturated voting card lay in the back of a white Toyota pick-up, its windscreen caved in as if hit by a heavy weapon. There seemed no plausible reason why election materials should have been found in the governor's compound. It wasn't much of a surprise when, in the presidential elections the following day, Alamieyeseigha's party was announced as the winner with 96 per cent of the vote.

I gathered further disturbing insights into Alamieyeseigha's style of government from – of all people – Freston Akpor, his spokesperson. Akpor said the governor travelled the state's rivers in a 16-capacity Sunseeker yacht bought with public money. These boats are – as the company website sales pitch has it – 'the Absolute in Performance, Style and Luxury'. Unfortunately, the Sunseeker wasn't exactly the most practical choice for Bayelsa's complex mosaic of waterways: Akpor said it couldn't fit down some of the smaller creeks.

Bayelsa State budgets suggested the Sunseeker wasn't the only example of opulence in the Alamieyeseigha administration. Between 2002 and 2005, the state spent more than $25m on his official mansion, the *New York Times* reported after studying official state documents. The state Poverty Eradication Committee received a little more than half of what was spent on toiletries for state officials.

Akpor told me that the governor had taken time out during his first term to study for a doctorate in strategic management from the University of Northern Washington. According to its website, the university is a Hawaii-based institution that offers distance learning courses and 'focuses on preparing graduates for a career in business administration'. If that seemed tangential to ruling a Nigerian state, then it was in keeping with the eclectic career history and qualifications Alamieyeseigha claimed. A former air force officer and company executive, his CV says he is also a member of Britain's Institute of Purchasing and Supply and an honorary professor at the Institute of Economics and Advanced Technologies in Ukraine.

Talking to other Bayelsans, I was not surprised to find out that Alamieyeseigha was a contentious figure whose rule was under heavy attack. His critics were surprisingly numerous, given the near-unanimous mandates he was declared to have won in elections. An Ijaw campaign group even passed me a dossier of allegations, bound in hard green covers, entitled 'Governor DSP Alamieyeseigha's Reign of Terror and Looting'.

The alarm at his behaviour was growing because of the swelling economic power of the governors of oil-producing states, which received an enhanced share of national oil revenues under a supreme court ruling. As the world oil price climbed, Bayelsa's takings topped £30m a month, more than enough to run a decent-sized district hospital in Britain. Just as the ebbing of centralized military control gave Lagos transport officials freer rein to take bribes, so politicians at the state and local levels had more power under civilian rule to run their territories autonomously and unaccountably. It was another warped – and counter-intuitive – democratic dividend.

Despite all the questions facing Alamieyeseigha back home, lawyers acting for the Nigerian government claim that he had little difficulty in using the Western financial system to serve his project of self-enrichment. Over his six and a half years in office, he allegedly racked up assets abroad of more than $20m, almost 10,000 times the earnings of one of his constituents living in dollar-a-day poverty. About $11.5m of this was sunk in four properties in London – including two in Cricklewood and one in Finchley – and a Cape Town harbour penthouse. Much of the rest was allegedly scattered in bank accounts controlled by him and his wife, at institutions including Barclays, HSBC, Royal Bank of Scotland and UBS. Other bank accounts were in the USA, Cyprus, and Denmark's Jyske Bank. A Jyske Bank brochure – slogan: 'Quite exciting. Quite safe' – lies on the floor of the Water Gardens flat's master bedroom.

The boldness of Alamieyeseigha's alleged corruption is clear from a successful damages claim launched by the Nigerian government to recover money from him. An April 2006 witness statement prepared by Colin Joseph, a London-based attorney for the government, claims that in 2001 the governor paid £100,000 of Bayelsa money in equal tranches into five bank accounts in the name of each of his five children. Nigeria's Economic and Financial Crimes Commission says these accounts had a staggering total turnover of £2.82m. Joseph says this is evidence of the 'remarkably blatant manner in which Alamieyeseigha distributed public money to his family'. The former governor has declined to respond to this claim or other allegations I put to him via his London-based lawyer.

It's striking how undeterred the banks or estate agents involved in the case were by the many warning signs surrounding their wealthy Deltan customer. Alamieyeseigha came festooned with what anti-money-laundering specialists call 'red flags'. One was that he was holding foreign bank accounts as governor even though Nigerian law did not allow him to do so. Another was that many of his assets were held through offshore companies based in secretive tax havens, such as the British Virgin Islands, the Seychelles and the Bahamas. A third was that – according to a declaration made when he took office in May 1999 – he had assets totalling $575,000 and an annual gubernatorial salary of just $12,000 a year. The numbers just didn't add up: the London UBS accounts alone, for example, had a balance of more than $2.3m by December 2005.

Documents assembled for the Nigerian government's London court cases to recover money from Alamieyeseigha show in particular the embarrassing eagerness of UBS to win the governor's business. The relationship dated back to 1999, shortly after he took office. An assessment made by the bank in December 1999 – six months after he came to power – gave Alamieyeseigha a clean bill of health, even though he was predicting a leap in his deposits from the existing $35,000 to $1.5m. The report claimed Alamieyeseigha was a financier and fertilizer magnate whose wealth 'predates his election and is clearly unrelated to his political activities'. The bank's public-spirited client 'does not intend to remain in office more than a few years, but has answered his people's call to steer the Bayelsa State through this difficult period of time'. This was the first time he had held a bank account outside Nigeria, it said, but he felt it was necessary to 'safeguard his substantial wealth from potential aggressors, for his own and his family's good'. All going well, UBS concluded, it 'intended to form a Trust relationship for this client'.

In April and May 2001, two years after Alamieyeseigha became governor, he suddenly injected $1.5m into his UBS account. Nasim Ahmed, a bank official who handled many of the dealings with the governor, prepared an internal report about how he had 'politely' asked the governor about the source of the funds. Alamieyeseigha told Ahmed that he had sold a palace he had built in Abuja to a Nigerian oil businessman and was expecting imminently a further payment of $1m relating to the deal. The governor said he expected the account to receive between $2m and $3m more from further property sales in Nigeria, to build up funds to pay for the education of three of his children in Britain and two in California. Ahmed pounced on this potential business opportunity. 'I tried to talk him into setting up a trust as these funds were meant as a nest egg,' he wrote. 'We spent over an hour explaining to him the structure and benefits. He finally agreed to our proposal to form a revocable trust.'

UBS London then prepared a plan for the trust to be set up by the bank's Bahamas operation. The document reported Alamieyeseigha's conventional biography at face value, with no apparent scepticism. The bank claimed 'Mr Alamieyeseigha commanded a lot of respect' and was thus 'elected to governor by the government'. The reality that Nigerian governors are elected and not appointed appears to have passed UBS by, which is perhaps just as well considering the suspected ballot fraud underpinning their client's success at the polls. UBS's Bahamas branch was more than satisfied with the result, however, complimenting its London counterparts for providing 'excellent information that helps us better know our client'.

The relationship between Alamieyeseigha and UBS continued to prosper, despite the growing number of questions being raised both in Nigeria and outside about the governor's conduct. In a May 2003 exchange of e-mails, two bank officials discussed what action to take about an item in the Nigerian media the previous month that suggested their client had been taken to court at home for allegedly breaking rules preventing senior elected officials from holding foreign bank accounts. One of the officials, Thierry Martin, dismissed the piece, saying press articles were 'not hard facts and should always be read between the lines'. The e-mail ended with an observation that suggested UBS took a pragmatic approach to doing business with foreign public officials accused of breaking their home nations' rules. 'Last but not least,' it said, 'if operating foreign bank accounts out of any country without reporting them to the relevant authorities was a crime . . . do I need to be more specific?'

Martin was the recipient of an equally troubling e-mail sent the following month, this time by Ahmed. In it, he described how Alamieyeseigha visited him to discuss buying a £1.8m central London penthouse through the UBS trust. Ahmed – seemingly more sceptical now – described asking the governor to 'put down in black and white with supporting documentation proof [of] the genuineness and sources of these funds'. Yet, even this demand was softened by the opportunity Ahmed gave the governor to go and think about his response: 'He went away and promised to come back with a plausible explanation about the sources of such funds,' Ahmed wrote. In the end Alamieyeseigha made the purchase through another financial institution.

UBS says it complied with all British money-laundering rules and co-operated fully with investigating authorities. Alamieyeseigha is no longer a customer and his money has been returned to the Nigerian government.

Alamieyeseigha was arrested by British police in September 2005 and charged with money-laundering. Although the reason why Scotland Yard began investigating him has never been officially confirmed, it's likely that one of the banks finally became too nervous and tipped off the authorities. Alamieyeseigha was remanded in custody initially but released after three weeks, on condition he surrender his passport and pay £1.3m bail. He then launched a counter-claim that he was immune from prosecution because he was a state governor, but his argument was thrown out by both Southwark Crown Court and the Court of Appeal.

In November, facing the inevitability of trial in Britain, Alamieyeseigha skipped bail and went back to Nigeria. Officials at Nigeria's Economic and Financial Crimes Commission said he had fled London disguised as a woman (he later described this claim as an 'insult' to a 'proud Ijaw man'). Nigerian news magazines had great fun mocking up cover photomontages of the governor dressed up rather fetchingly in robes and a shawl. His bail was forfeited; in a piquant twist, one of the associates who had offered to put up the money, Terry Waya, was himself investigated by police and later convicted of mortgage fraud.

Having made it back to Nigeria, Alamieyeseigha probably thought he was safe, thanks to a much-criticized clause in the constitution that gave state governors immunity from prosecution. But, unfortunately for him, lawmakers in his state, under pressure from both Nigeria's central government and from overseas, voted in December 2005 to impeach him. He was charged with money-laundering and corruption by the Economic and Financial Crimes Commission. The EFCC claimed he had stolen cash, and also taken bribes from business people in his state who had government contracts to supply power generators and other goods. In July 2007, after the former governor admitted the charges relating to false asset declarations and money-laundering, the Nigerian courts ordered many of his domestic assets to be confiscated.

In follow-up judgements in the High Court in London, Alamieyeseigha was ordered to forfeit his four London properties and the money held in the UBS account. A High Court judge had noted in 2007 that the former governor, who offered only the sketchiest of defences throughout, had 'a lot of explaining to do'. In December that year, Mr Justice Morgan said Alamieyeseigha had now 'had the opportunity to face his accusers' but had failed to counter their claims.

'He had that opportunity in the Nigerian court,' the judge said. 'The charges against his companies would have had to be proved there beyond reasonable doubt. He has not felt able to challenge the allegations against the companies and, because he controlled the companies, these are allegations which are in effect against him also.'

Alamieyeseigha's case was far from the only one to suggest that the proceeds of Nigeria's oil bonanza were being diverted in deplorable and extravagant ways. Scotland Yard alone had its eye on two other governors, one of whom, Joshua Dariye of Plateau State, also skipped bail after being arrested in London on suspicion of money-laundering. The English High Court has since ordered Dariye and his associates to pay more than £4m of suspected stolen money back to the Nigerian government. Another oil region governor, James Ibori of Delta State, had his British assets frozen on suspicion that he had laundered at least £30m through the country between 2005 and the end of his term in 2007. English High Court papers show that Scotland Yard investigators suspected corruption in a deal struck by a company controlled by him, MER Engineering, to supply Shell and Chevron with houseboats on which oil workers could live (the multinationals declined to comment, while Ibori has denied all wrong-doing). By the end of Obasanjo's second presidential term in 2007, the Economic and Financial Crimes Commission said it was investigating more than two-thirds of the country's state governors, as well as many other senior officials.

My interest in cases like those of Alamieyeseigha and Ibori was part of a growing – some would say obsessional – fascination with the tentacles of Nigerian oil-related suspected corruption and how they stretched from the local to the global. The longer I lived in the country, the more I became intrigued by the psychology and consequences of graft at all levels of society. During my early days in Lagos, I used to joke and smile with the police road checkpoint officers who asked me for bribes, until an expatriate financier – the irony! – scolded me for abetting crime. As a rich foreigner, he argued, I was in a privileged position to try to stop it happening. I could speak out without fear of being beaten up, unlike the many poor Nigerians from whom the police daily extorted cash. If I said nothing – or, worse still, gave the impression that I thought the requests for money were OK – I bore the same responsibility as banks that turn a blind eye to money-laundering. Better to sound like a pompous arse, he said, than to be complicit in oppression.

Persuaded by his analysis – and not being a stranger, as friends have pointed out, to pomposity, piousness or puritanism – I tried a change of strategy. From now on, I decided, I would confront all officials who asked for bribes. The results were, to put it mildly, mixed. Sometimes it worked, and the police would back off, perhaps afraid that I would report them to influential people I knew. At other times, especially late at night, they'd had too much beer or palm wine to register what was going on, leading to farcical dialogues of the drunk.

The best response of all, which picked out the absurdity and borderline distastefulness of my lectures to people who earned no more than a few per cent of my salary, came when I was on my way to a nightclub near the airport. The officer who stopped me at the checkpoint asked me, in time-honoured fashion, whether I would consider doing something for him for the weekend. 'Are you asking me for a bribe, officer?' I said. He looked a bit puzzled and thought for a moment. 'Yes!' he exclaimed, apparently mystified at my failure to take the hint.

Yet, beneath the ridiculousness, I was glad I never did pay. Corruption had come to seem to me one of the most serious problems a society can face. At its worst, it condemns people to poverty and even death, because resources have been stolen or projects left unfinished. Bribery is also – even by the debased standards of global politics – one of the biggest sources of international hypocrisy. Almost every government officially agrees it's a bad thing, but almost none is interested in taking serious action against it. That double-think unites capitals between whom there is otherwise little love lost, from Moscow to London and from Washington to Beijing. The USA has done more than most countries to combat bribery over the past five or ten years, but even American investigators have barely scratched the surface of the corruption that so thickly coats international finance and the oil industry in particular.

The grand-daddy of all Nigerian corruption cases, which had been rumbling for years before I arrived in the country and continues to this day, was the international pursuit of money allegedly looted by the late dictator General Abacha and his associates. This epic, ongoing hunt has involved billions of dollars and has embarrassed banks, companies and governments across the world. Foreign countries have enjoyed a grotesque double benefit from the Abacha theft. Not only did they buy Nigerian crude, but their banks took a hefty portion of the proceeds from the sale of it. It is yet another thread linking the wider world with Nigeria's continuing failure to register the development surge the world's oil century ought to have allowed it to enjoy. The scale and importance of the Abacha theft is evoked by Enrico Monfrini, the lawyer charged with recovering the stolen assets for the Nigerian government. 'Under the dictatorship of General Sani Abacha,' he wrote in his original court deposition in 2000, 'corruption reached heights that were previously unknown in the history of Nigeria, and rarely seen in the history of the modern world.'

Abacha was the last of three successive military rulers whose joint tenure spanned a decade and a half, save for a short-lived period of civilian transitional government. In hindsight, those years of ever-increasing brutality and crudeness resemble stories of dictatorial decline and decadence from the Roman Empire and beyond. Since Abacha died in 1998, there has been a tendency among many outsiders to see his regime as uniquely wicked, even by the standards of the country's violent and corrupt politics. But quite a number of Nigerians see this as a mistake that lets other leaders off the hook. For many Nigerians, Abacha is not a one-off, but rather a roughly hewn archetype for the behaviour of the political élite and their Western accomplices before and since, as hundreds of billions of dollars' worth of oil revenues have been squandered. Under Abacha's predecessor, General Ibrahim Babangida, for example, more than $12bn of revenue windfalls generated by the spike in the oil price during the first Gulf War were never properly accounted for. A later investigation into the fate of the money was never published. The charming, gap-toothed general – who enjoyed then prime minister Margaret Thatcher's hospitality at Downing Street – earned himself a set of picaresque nicknames back home. Some of these – principally the 'evil genius' and the 'Maradona of Nigeria' – suggest the curious mixture of loathing and awe many Nigerians have felt for their cannier overlords.

One reason Abacha comes in for such opprobrium is that he ruled with open brutality compared to the craftier repression of some of his predecessors. In short, his public relations were lousy compared with theirs. During his five-year dictatorship, his name became synonymous with the kind of open corruption and oppression that is the product of untrammelled power. He locked up political critics and rivals, including the one-time dictator and future civilian president Olusegun Obasanjo. The state became increasingly militarized, with army checkpoints on the roads. Human rights activists, journalists and other citizens were harassed, beaten and sometimes killed. Increasingly cut off from the outside world – apart from the oil industry, which continued to generate big profits for both multi-nationals and the regime – Abacha and his cronies did pretty well as they pleased back home.

For Nigerians suffocating in this ever-tightening embrace, a moment of merciful release finally came with the general's sudden death on 8 June 1998. A few days later, as if in cathartic celebration of this so-called 'coup from heaven', Nigeria scored a famous victory over Spain in the football World Cup. Nigerians' reactions to Abacha's death revealed the bleak places to which the regime and its predecessors had taken the country. 'It was perhaps the first time in my life that I remember being glad that someone had died,' recalls a Nigerian-born acquaintance of mine who now lives in the USA. 'Frankly, I felt some shame for feeling as I did.'

Once Nigeria returned to civilian rule in 1999, the new government hired the Swiss lawyer Enrico Monfrini to try to untangle the webs of ancien régime corruption. The nerve-centre of his efforts is Geneva, where his law firm, Monfrini Crettol & Partners, specializes in the international recovery of stolen assets. Approaching his base one winter's day, I quickly see its ambience is far from the everyday mayhem of the country he is investigating: his office is a short step from the lake shore, down a corridor of shops that include Cartier and Louis Vuitton. The building overlooks an attractive square, where the late afternoon sun warms the cobbles, pigeons and shoppers.

Monfrini himself appears to greet me at the door of his second-floor office. We'd met briefly the day before, when I'd been struck by the juxtaposition of his slick silver hair with a weather-beaten face that suggested prolonged exposure to the African sun. A deep scar runs down from the top of his forehead, a valley wending through crags of flesh. He is not physically imposing, but he has a certain authority and gravitas that he leavens with dry asides. On our first meeting, he had produced a part-drunk large bottle of Volvic mineral water, which he emptied at a single draught. 'It's good for the health,' he'd explained in French, adding with a chuckle, 'They say whisky is better.'

In Monfrini's private office, a single patch of sunlight in the far corner catches the shadows of the railings outside, casting prison-like bars across the floor. The shelves, bare of papers, are dominated instead by a collection of antiques. To my left are statues and masks from West and Central Africa, including a Nigerian terracotta figure of the Nok civilization that – if genuine – must be more than 1,500 years old. In front of me, adjacent to his desk, are two Catholic icons, with the Virgin Mary picked out in gold leaf. Most eerily of all, to my right, in my eye-line as I speak, are about 20 Chinese Han Dynasty figures, standing and turned to face in my direction. 'They are looking at you,' Monfrini says, when I remark on the discomfiting effect. 'But they are also protecting you.'

This playful observation is typical of a raconteur who has a theatrical sense of the props around him, notably his Lino-brand cigarillos. He toys with them, picks at them and smokes them throughout our conversation, even though the packet tells him in three languages that they will kill him in the end.

As he leans back and puffs, he reminisces about how his long-standing personal links with the African continent helped lead him to the Abacha job. His family roots are in Milan, from where his great-grandfather emigrated as a vagabond when Italy was united in the nineteenth century. He became rich and handed on the money to Monfrini's dissolute grandfather, who spent so much time 'hunting and chasing women and animals' that he died almost bankrupt. Monfrini's father rebuilt the family's standing, first as a lawyer and then as a diplomat in Ivory Coast, Gabon and Ghana. The teenage Enrico used to visit him and travel around the region, a young man on an African adventure. 'It was,' says Monfrini, 'a lot of fun. AIDS didn't exist at that time.'

Monfrini declares – with a surprising bluntness – that he likes black people and knows 'how to deal with them'. Africans appreciate his straightforwardness. Those he knows tend not to be hypocrites, and are usually prepared to talk a problem over. 'If something doesn't go, I tell them,' he says. 'I don't try to pretend to be the good white man. I don't pretend to be totally non-racist either. I pretend to be what I am.'

The night Monfrini first heard about the Abacha job, he was at Geneva's pricey Lipp restaurant. He'd just eaten oysters and a big plate of *choucroute* – cabbage and sausages – washed down with a 'marvellous Sancerre'. Then, just before midnight, his mobile rang. On the other end of the line was a stranger who said he was a member of the Nigerian government. 'He said he actually was the national security adviser of the country, and that he wanted to see me,' Monfrini recalls. 'So I said, "When?" He told me, "Now."'

Monfrini jumped into a taxi and headed for the Hotel du Rhône, where rooms were advertised in early 2009 at upwards of £375 a night. 'I was not 100 per cent sober,' Monfrini recalls. 'But I was OK.' At the hotel, he found General Mohammed Aliyu Gusau waiting for him. Gusau was businesslike and to the point. He said the new Nigerian civilian government wanted to hire Monfrini to track down money it suspected had been looted by the recently deceased Abacha. The work was urgent, so Monfrini needed to come to Nigeria as soon as possible to start it. Gusau suggested leaving the following day, Monfrini recalls, with another rumbling smokers' chuckle. 'So I said, "OK: I'll come in one week."'

Monfrini remembers that he 'smelled Africa again' when he arrived in Nigeria to start work on the Abacha case. It's not always easy to live on the African continent, but it is 'extremely pleasant'. There's a feeling of 'strength, of grandeur, of eternity', which you don't find elsewhere. He talks, with Conradian echoes, of Africa's wildness, its sense of the unknown and its 'lack of history'. If his terracotta – a sculpture possibly older than Islam – could talk, it might have interjected at this point.

The lawyer says he doesn't know exactly why he was chosen for the Abacha assignment, although he does have some ideas. He had worked over the years for Nigerians in the USA and Europe who were opposed to Abacha's regime. 'I had this network over so many years and they acquired trust in me,' he says. 'You know, the African approach is that once you are trusted you are trusted, and they trust you with everything they have.'

He recalls how his first visit to Nigeria showed the forbidding size of the task he faced. The Nigerian authorities had only a handful of documents relating to the graft. They were sparse but vital leads in the chase to come. 'I spent four or five days in Nigeria speaking with these guys, trying to collect what I could,' Monfrini says. 'And I left

Nigeria with maybe 30 pages of documents, which was actually not much to start this battle.'

Other lawyers involved in the Abacha case speak highly of a preliminary investigation carried out by Peter Gana, an assistant commissioner in the Nigerian police. Tim Daniel and James Maton of Edwards, Angell, Palmer & Dodge, which acted for the Nigerian government on the London side of the hunt for the suspected stolen money, have described how Gana 'amassed a significant amount of evidence'. He identified some of the methods used by Abacha to extract cash from the government, particularly 'the removal of massive sums from the Central Bank of Nigeria under the pretext of being required for national security purposes'.

The most intriguing feature of the original Nigerian documents, Monfrini says, was the leads they gave to money movements involving Swiss banks and foreign banks operating in Switzerland. In some cases the name of the bank was given; in other cases only the account number was available. Monfrini decided to use these details to draft a request for legal assistance to the Swiss authorities. He describes the letter now as a kind of trompe l'oeil, designed to persuade the Swiss that the Nigerian government had much stronger evidence than it actually did. Monfrini calls the application a 'masterpiece' of presentation that suggested the case was 'completely clear and extremely obvious'. 'The name of the game,' he says, 'was to put a lot of condiments in the letter so that the Swiss authorities would bite.'

Within two weeks, the Swiss had swallowed the bait. The authorities – under pressure to clean up after the Nazi gold scandal and other problems – froze a number of accounts named by Monfrini. The action 'burst like a bomb in the banking world in Switzerland', but its immediate practical impact was small. The banks' records showed the accounts were empty, suggesting the money had gone elsewhere.

Monfrini decided to try another, harder tack. He launched a legal complaint in Switzerland in which he used the device – then novel in international corruption cases – of accusing Abacha and his associates of acting as a criminal organization. This approach, more often employed against drug cartels than venal heads of state, raised the stakes in the case. Most importantly, it meant the onus was now on those suspected of laundering the money to prove it was honestly acquired. The Swiss authorities sent out a blanket alert to hundreds of banks asking if they held any accounts whose names and owners corresponded to those on a list provided by Monfrini. That, the lawyer says, triggered a 'monstrous discovery': almost $700m of Abacha money was in the Swiss banking system.

Once the money broke the surface of Swiss financial secrecy, banks started sending the authorities reams of documents linked to the suspect accounts. Monfrini recalls the papers piling up in his office. He had five people poring over them night and day, trying to work out how the money had flowed around the world. He and his colleagues survived on 'sandwiches – and sometimes Sancerre', Monfrini says, with another growling rumble of amusement.

The work was complex, for reasons wearisomely familiar to international anti-corruption investigators and campaigners. One was that the suspected money-laundering appeared to involve at least half a dozen countries on three continents. A second difficulty was that the currency in which the money was held often changed as it shifted from nation to nation, making the movements hard to follow. A third problem was that money from different accounts was constantly being blended, altering the sums involved. Monfrini describes the analysis as 'monstrous work'. 'Our job was not just to be lawyers, but to become more like accountants,' he says.

He knew he was still only at the first stage in a punishing process. By early 2001, when he had a pretty good idea of how the suspected money-laundering worked, political obstacles had already started to appear. The next phase would be attritional, expensive and unending. As Monfrini puts it, 'It took less than one year to identify all the funds we have now been struggling for for eight years.'

One of the biggest problems Monfrini faced was following the many money trails that led to and from Britain. While other states were willingly providing information, Britain was proving singularly recalcitrant. It took years to respond to a request for bank data and then provided documents that were 'mostly useless'. Monfrini says that, 'for funny reasons, somebody decided – I don't know who – not to do the job'. Asked what he means by this, the otherwise loquacious lawyer will say no more.

London mattered because of its crucial role as a trans-shipment point for the Abacha money. In 2001, Britain's Financial Services Authority estimated that at least $1.3bn of the loot had passed through 23 London financial institutions. London's failure to do more to help Monfrini's search after this disclosure is much criticized in Nigeria and acknowledged privately in Whitehall. One foreign office official admits there is a 'sense of embarrassment' in London about the Abacha case, especially given the progress in Switzerland. Another official says the British government was 'too eager to throw requests back' to the Nigerians on technical grounds, rather than advising on how to improve them. This foot-dragging is an eye-catching counterpoint to the boasts by former and current British prime ministers Tony Blair and Gordon Brown of their commitment to combating corruption and promoting development in Africa.

Over the years, Monfrini chipped away at the case in Switzerland and elsewhere, freezing money as he found it and pursuing the long process of seizure. Notable triumphs include $500m sent back by Switzerland since 2004 and $160m returned by Jersey in 2003. As of early 2009, the tally of recovered Abacha loot stood at about $2bn. Of this, authorities in Nigeria seized $800m in the aftermath of the fall of the Abacha regime. Monfrini has recovered almost all the rest.

Lawyers in London clawed back a further £110m in a court case that came to symbolise how the suspected Abacha corruption thrived in the darker corners of the international financial system. The lawsuit centred on allegations that the general's associates had illegally taken a large cut from the proceeds of complex transactions involving debt owed by Nigeria to Russia. The loans in question had helped to finance the Ajaokuta steelworks, which consumed billions of dollars in investments from the 1970s onwards, but was virtually defunct until attempts were made to revive it in 2005. The English High Court found in 2001 that a British Virgin Islands company controlled by Mohammed Abacha, one of the general's sons, and Abubakar Atiku Bagudu, one of the late dictator's advisers, had been used to skim off profits made through buying Ajaokuta debt from the

Russians and then selling it on to the Nigerian government. During the trial, the court took evidence from both Bagudu and Mohammed Abacha, who was in jail in Nigeria over a suspected murder case that was later dropped. In his final judgement, Mr Justice Rix said Mohammed Abacha and Bagudu were 'unreliable' witnesses, adding that Abacha was 'dishonest' and Bagudu 'capable of dishonesty'.

Few people outside the Abacha family would dispute Monfrini's claim that there should be still more money to come back to Nigeria. By 2009, he had identified further targets amounting to more than $1bn, much of it in tax havens. He had decided to focus on $300m in Jersey, $300m in Liechtenstein, $350m in Luxembourg and $60m in the Bahamas. Then there are funds that have yet to be tracked down, because they are held in cash or are sitting unnoticed in accounts fronted by friends of the regime. More than a decade after he was hired to investigate the Abacha case, Monfrini was still pursuing legal action in Austria, the Bahamas, the Cayman Islands, France, Germany, Kenya and the USA.

Part of the reason Monfrini has been able to recover so much so far is that Abacha's methods were so crude. According to a Nigerian investigation, much of the corruption was a straightforward theft of almost $2bn in cash and travellers' cheques from the country's central bank. Some of this cash was simply taken physically to overseas banks. Monfrini's first court deposition claims that 'significant quantities' ended up in Britain. One of Abacha's most active bagmen was Ahmadu Daura, who owned a foreign exchange bureau in Nigeria called, with incongruous innocence, 'Sunshine'. In 1998, Daura was arrested at Heathrow Airport, after stepping off a private jet with £3m stuffed in his suitcases. 'It is believed,' Monfrini's court deposition notes drily, 'that Mr Ahmadu Daura was not at his first attempt to bring large amounts of cash to the United Kingdom.'

The Nigerian government's claims suggest that the Abacha looting was perpetrated on an industrial scale to rival the Mafia. Some of the biggest suspected clearing houses for the stolen money were bank accounts opened by Mohammed and another of General Abacha's sons, Ibrahim, under the codenames Kaiser and Seuze. As one journalist has pointed out, the choice of account names brings to mind the mysterious figure of Keyser Söze, the spectral gang boss in the film *The Usual Suspects*.

Nigeria's government claims that some of the overseas bank accounts were controlled directly by Abacha and his associates, and others by Nigerian or foreign business people who passed on the proceeds after keeping a commission. One vehicle allegedly used was the Nigerian Family Support Programme, which imported vaccines. The agency is suspected of giving a $111m contract for the supply of vaccines to Morgan Procurement, a company controlled by Abacha associates. The vaccines allegedly cost just $22.5m, leaving the suspected looters with a cool 400 per cent profit.

The role of international banks in facilitating these kinds of transactions has become something of a cause célèbre, showing how the venality of the Western financial system has helped sustain dictatorship around the world. Most notably, in 1999, the US Senate Permanent Subcommittee on Investigations issued a stinging report on Citibank's conduct in the Abacha affair. The document is a classic, relentless account of how a financial institution was more interested in holding on to its clients' money than in asking too many questions about where it came from.

Abacha's sons Mohammed and Ibrahim became Citibank clients in 1988, five years before their father took power. When the brothers opened accounts in New York in 1992, Michael Mathews, who was responsible for their London holdings, sent a reference in which he described them as 'the son and adopted son of Zachary Abacha, a well-connected and respected member of the Northern Nigerian community.' At this time Abacha was minister of defence in the military junta that preceded his own.

Mathews highlighted why he believed the sons to be good and potentially lucrative clients. The sons were operating the accounts on behalf of Abacha senior, making them 'clearly target market by association'. The accounts were used to hold the proceeds of commodity trading and had been operating 'entirely satisfactorily', although their balances had 'fluctuated wildly'. The sons themselves were 'unfailingly charming, polite and, above all, reliable', in contrast with other Nigerians Mathews dealt with.

One of the bank's internal documents described Mohammed Abacha, who was operating under the less than subtle false name of Mohammed Sani, as 'son of minister for youth and sport'. By this time, Abacha had been military dictator for more than two years. Only in 1997 were the bank's documents updated to read: 'Muhamed [sic] Sani, Son of Nigerian President . . . Account opened several years before father became President.' The Senate subcommittee report describes the Citibank New York's lack of knowledge of who its clients' father was as a 'critical lapse in due diligence'.

On one extraordinary occasion, Citibank even acknowledged the suspicions surrounding the origins of the very money it was banking. Its 1997 client profile relating to one of the New York accounts described Abacha's wealth as coming from his links to the state Nigerian National Petroleum Corporation. Lower down, with breathtaking insouciance, the bank recorded that the general was the current military ruler of Nigeria, 'where there is a lot of corruption in connection with the petroleum industry'. Despite this, Citibank took comfort because it had always found the Abacha sons 'good and very professional'.

In 1999, Citibank finally decided to close the accounts after a review of its policy on clients who were public figures. Before it could do so, a London court froze money relating to the Abacha organization held at several London banks, including Citibank. It was the end of a sorry episode during which, the senate investigation noted, Citibank's conduct was characterized more by 'customer deference' than due diligence.

Citibank was far from the only company to become embroiled in the investigations. The list of banks that Monfrini claims processed Abacha-related money reads like a *Who's Who* of Western financial institutions. In 2001, the High Court in London issued freezing orders against accounts at Britain's HSBC, Barclays and NatWest; Merrill Lynch International Bank and Citibank of the USA; Deutsche Bank and Commerzbank of Germany; BNP Paribas and Crédit Agricole Indosuez of France; and Credit Suisse First Boston and Union Bancaire Privée of Switzerland. Some of the institutions that were named were claimed to have only tangential links to the case, but the sheer number of them named gives a sense at how the late dictator's money spread like an oil slick right through the international financial system.

A further significant strand of the Abacha corruption is its alleged links to leading international companies outside the banking sector. This is outlined in an unpublished Swiss Federal Office of Justice probe that is an eye-opening complement to the US Senate investigation. Several well-known businesses wanted to win the contracts funded by Nigeria's oil wealth. As the Swiss report concludes, many of them seemed prepared to pay substantial bribes to the Abacha organization to ensure a share of the spoils.

Flicking through the report, it is clear that Abacha-era corruption in Nigeria was a cosmopolitan phenomenon. Like any successful multinational, the Abacha organization had a truly global reach, taking in client businesses from Western countries, China and India. As with the passengers on Agatha Christie's Orient Express, everyone was in on the crime. It seems most of the world was on the make or on the take, whether by paying bribes or by ignoring what was going on.

One transaction the Swiss report focuses on involves the German telecommunications multinational Siemens, which has since been engulfed publicly in a scandal over suspected corruption. The company had several contracts in Nigeria, including one for laying fibre-optic cables for the state telephone company, Nitel. According to Monfrini, three payments totalling more than £7m were made by a middleman acting for Siemens into accounts controlled by the Abacha organization. The Swiss Federal Office of Justice says it could find no economic reason for the payment into the private account of Abacha's sons. Siemens has declined to respond to my questions about the case. In December 2008, the company agreed to pay almost £1bn in fines to US and German authorities over suspected bribery around the world, including in Nigeria.

The Swiss investigators were to receive no enlightenment from Abacha's son Mohammed, or a third son, Abba. The report notes that they 'failed in their duty of co-operation' with the authorities, by not giving any explanation of the origins of 'the considerable sums deposited in their bank accounts'. The document concludes that there is 'every reason' to believe the accounts held for and on behalf of 'the Abacha clan' in Switzerland were 'intended exclusively to receive the proceeds of the crime of which they are accused'.

The Swiss report is a sobering insight into the realpolitik considerations that govern international attitudes to corruption, for all the condemnations made of it in recent years. Reading through the document leaves me with a frustrating awareness of unfinished business: dubious transactions and suspicions never quite confirmed keep cropping up. For all the reams written about the Abacha case, it is clear that the full story has never emerged and may never do so.

One reason for this, as the Swiss report makes clear, is that the Nigerian government has never disclosed all it knows. Nigerians are unsurprised by this, as the case has the potential to embarrass too many people who are still in public office under civilian rule. The government will release what information it needs to in order to get the money back, but no more. As one Nigerian lawyer friend puts it: 'Much as the Nigerian government wanted the Abacha money back, they didn't want the full story to come out.' It helps that Switzerland, under fire for its financial secrecy, wants to be seen to be bending over backwards to help.

Another intriguing lacuna, Monfrini suggests, is that none of the money so far found appears to come from international companies in the oil sector. It is, he says, taking a long puff on his cigarillo, 'very amazing' to see that 'not a cent' of the $2bn he has identified as corrupt payments to the Abacha organization comes directly from the oil business. It is 'extremely difficult' to imagine that the Abacha family did not make any money out of the oil industry.

Given the big bucks involved, it's no surprise to hear Monfrini say he received death threats during his investigation. The warnings came via lawyer and banker friends, who told him vaguely about 'dangerous' people who were 'very angry' with him. He was told to be very careful. 'So I said, "What do you mean, very careful? Is he going to kidnap my children, kidnap me, or rape my wife or what?" So you see, these kind of subdued threats came very often,' Monfrini says.

Monfrini says he has also had many offers of bribes to stop his work. He would, he tells me with feeling, have been 'a multi, multi,

multi, multi, multi-millionaire by now' if he had taken up the proposals that came his way. They came from agents for bankers and agents for Abacha associates. 'I could have taken so many bribes I would be very *reech*,' he says, stressing the last word. 'But I'm not of the kind. I don't want to say that I am the best and I don't want to say that I am the most fantastic. But in terms of trust I don't like to fool my clients.'

I ask Monfrini to tell me a bit more about the Nigerian clients for whom he acted before the Abacha case. He won't name any of them, although he says they are all 'more or less wealthy' business people. Unlike the newly minted Abachas, Monfrini's clients mostly inherited family money.

'There's also a lot of old money in Nigeria,' he says, 'which might be also the proceeds of crimes, but very old crimes.'

I am surprised at his speculation about his clients' backgrounds. I ask whether he was ever worried that he was being paid in money itself stolen from the cauldron of crude riches. 'No,' he replies, 'because my clients do not steal. Maybe their fathers did.' He lets out another rumbling chuckle. Is he comfortable taking their money, though? He is unusually hesitant, taking a while to find the words he wants. He says: 'It's . . . er . . . it's er . . . legally OK and morally, I think . . . acceptable.' What would I do, he retorts, if I inherited family money that I knew was made through my grandfather selling alcohol during the time of prohibition? 'Are you going to refuse this money?' he asks. 'How far is the moral reservation supposed to go?'

His wife would take a hard line on this, he says, with another swell of laughter. She has no illusions and is 'very concrete and very Catholic'. She loathes dishonesty and anything connected to tainted money. 'And she hates people with money,' he adds thoughtfully. 'So she hates me,' he concludes, releasing a peal of mirth that suggests his reply shouldn't be taken at face value.

Monfrini seems keen to continue this line of conversation, even though he doesn't appear very comfortable with it. His wife is unhappy about the fees he is earning from the Abacha case, he says, as she thinks they are the proceeds of crime. In a way she is right, he reflects, adding that he is sometimes troubled by the thought of how the restituted money will be used in Nigeria. In short, he is worried that this looted oil money that has taken so many years to recover could simply be stolen again.

A question that's even closer to home for Monfrini is whether his time-consuming and expensive work has given Nigeria value for money. He is coy about his own earnings from his work, admitting only that he has received millions of dollars.

'It's more than you think,' he says, 'but less than you imagine.' Is it more than $10m? 'I wouldn't say that,' he replies. Is it less than $10m? 'I wouldn't say that,' he repeats. 'It's something which I cannot really disclose,' he concludes, adding that the Nigerian authorities have constantly tried to renegotiate his bills.

Monfrini has no regrets about taking on the Abacha job, he says, although he admits to being 'rather innocent' when he started it. He was sure that, once he found the money, Abacha's associates would be pursued and prosecuted in Nigeria. Instead, they have walked free, cushioned still by some of their stolen money. In Nigeria, Monfrini says, he 'had no real support to deal with these Abacha people, because actually nobody dared'.

I wonder aloud if that means Abacha's suspected accomplices have, in the end, won the battle. Monfrini, not surprisingly, disagrees with that. He says their victory, if it is that, is Pyrrhic. They have their liberty and some of their money, but most now live in a gilded cage in Nigeria, unable to travel outside the country. 'They are not free, because they can go nowhere. They can't even use their money except to pay their servants and buy good food.' I see his point, but I also see that those alleged to have profited from a dictator's looting are very far from enduring real suffering.

Part of the reason Monfrini was able to recover as much money as he did is that the political circumstances in Nigeria surrounding the case were so unusually favourable. Abacha was dead and had a terrible reputation, so there were not many people in government who were prepared to fight on his behalf. The new president, Olusegun Obasanjo, had been jailed for several years by Abacha and had a score to settle. 'It didn't only have to do with corruption fighting,' Monfrini says. 'It had to do with revenge. So that is a major ingredient for success.'

Without that 'complete support' from the looted country, pursuing corrupt leaders in resource-rich nations such as Nigeria is very hard, Monfrini suggests. If international campaigns to return dictators' stolen assets are to be successful, much tougher action is needed. Governments will have to take 'forceful steps', he says, pushing his hands out like an icebreaker confronted by a glacier. In most cases, governments looted by kleptocrats have little interest in pursuing former rulers, as those newly in office usually owe political debts to their predecessors and may even have been hand-picked by them. Often, Monfrini says, it will be possible to provoke governments into action only by 'paying them, or by threatening them'. 'It's a very delicate issue,' he says.

Given all the obstacles stacked against successful investigations, Monfrini accepts that a case like the Abacha looting could happen again in Nigeria. Progress has been made in tackling corruption both in the country and internationally, he says, although there are still loopholes in rules, failures in banks and many safe financial havens. It is still 'completely possible' for present and future kleptocrats to stash stolen money around the world, he says, adding, 'But it will become more and more difficult for these people to do what they essentially need to do with the money they steal.'

As I prepare to leave, Monfrini tells me that he doesn't really see his case as a fight against Abacha and his suspected accomplices. Without being too heavy or philosophical, he says, he believes his work has changed the face of the world of international corruption 'a little bit, a tiny bit'. It is more than just another legal brief in a life during which he must have taken on hundreds, if not thousands. 'We do our jobs to make a living,' he says, 'but mostly because we want to achieve things, important things, as important as we can, to bring some truth into the system.'

Stepping out from Monfrini's office into the early evening winter chill, I see why the Abacha case was in important respects a great success. But I also see a plausible argument for finding it a troubling commentary on international anti-corruption efforts in Nigeria and other resource-rich countries. Here was a case in which, legally and politically speaking, the planets were aligned to help Nigeria recover the money. Yet, more than eight years on, Monfrini was still chasing the cash and many of the Abacha organization's members were living with impunity.

Nor have Western banks and companies suffered greatly for their role in the Abacha affair. Britain's Financial Services Authority said it had disciplined 15 banks, but they remained anonymous. The Swiss Federal Banking Commission studied 19 banks, finding that six failed to perform adequate checks on the money and 15 had lax anti-money-laundering controls. An extraordinary review was ordered and some bank directors were dismissed. A new corruption law and new rules on due diligence and money-laundering were introduced. But no one knows how much banks around the world earned in fees related to the Abacha loot, or what they did with much of that money.

Perhaps the story of the Abacha organization and its great oil-wealth grab should be seen in the same light as that of the nineteenth-century US robber barons, many of whom were deeply involved in the early days of drilling their country's crude. Ruthless and often unscrupulous, they amassed huge fortunes until changes in the law broke up their business empires. By then, their families were assured of perpetual power and praiseworthy endowments to charitable foundations. Viewed in this light – and that of the more recent cycle of Wall Street profiteering and subsequent chaos – the grand corruption of Abacha and his venal predecessors is not so far removed from the Western capitalist tradition as it might appear.

The difference is that, whereas the nineteenth-century US plutocrats left behind productive companies – some of which now export Nigeria's oil – the Abacha organization presided over the mass impoverishment of one of the world's most populous nations. The oil money simply disappeared from the country, with the willing assistance of some of the world's leading banks. Since civilian rule returned, no one knows how much money from Nigeria's record receipts from crude has flown to foreign accounts in the hands of officials like the former Bayelsa governor Alamieyeseigha.

The Alamieyeseigha case shows how, for all the euphoria surrounding Abacha's death and the recovery of some of the money he stole, Nigeria's circumstances were never likely to allow the fresh start its people desperately hoped for after dictatorship ended. Nor was it ever likely that the subsequent transition to civilian rule would magically kickstart a period of progressive government in which the country's oil resources would finally be harnessed for the wider public good. The government elected in 1999 did promise reforms that it said would, over time, help turn the country around. What happened proved a more nuanced and darker tale that left many observers of Nigeria more fearful than ever before for the future of the country, its crude and its relationship with the rest of the world. As the nation approached a decade of civilian rule, it found itself at the centre of a gathering international conflict that threatened to intensify once again the century-old war over the Delta's oil.

As I ploughed through the documents in the Abacha and Alamieyeseigha cases, I realized that I'd been wrong about the date of my first brush with the former Bayelsa governor. One of his Nigerian properties seized by the court in Lagos was Abuja's Chelsea Hotel, where I'd stayed in 2002, during my first week in the country. If it was named after the bohemian New York lodging celebrated in the Leonard Cohen song 'Chelsea Hotel', the choice seemed apt. Abuja, the capital and centre of government wealth, was where many politicians and business people came to do deals and be away from their wives back home. Not a few guests at the Chelsea and the myriad other hotels like it were, as Cohen put it, 'running for the money and the flesh'.

I remember the visit mainly because of what had happened when I checked out. I had only a fistful of British pounds, as I was a few days into my time in Nigeria and hadn't yet worked out how to tap into the network of 'Alhajis' who dominated the thriving industries of bureaux de change and roadside money swaps. I handed over £120 to pay the £117 bill, only to see a look of dismay cross the charming young receptionist's face when I asked for change.

'Well, you *could* pay £117,' he said doubtfully. 'But I know £3 means nothing to you.'

I looked at him and registered one of my first glimpses of what a colleague once described as a 'Nigerian twinkle'. The receptionist seemed to have accurately clocked me as a naïve JJC – Johnny Just Come – flush with foreign currency and unlikely to have the right money for anything. I realized there was only going to be one winner of this battle of wits.

'Take it,' I said, 'as a token of my gratitude to the Chelsea Hotel.'

The receptionist inclined his head graciously, almost apologetically. 'On behalf of the Chelsea Hotel, I say thank you,' he concluded, whisking away the £120 and banking another tidy little unearned profit for the Alamieyeseigha financial empire.

Part Three

THE NEW GULF CONFLICT

Dispatches from the Western Front of the Battle for Oil

'Many foreign companies dey Africa carry all our money go. . . International Thief Thief.'

— *lyrics from* International Thief Thief,
by Fela Anikulapo Kuti

7

FISH, BUT NOT FISHING

On the high seas of West Africa's Gulf of Guinea, the rear deck of the US warship *Dallas* is a nest of smokers and fishermen. The crew members are kicking back on this sunlit Sunday morning, enjoying a rare spot of 'holiday routine' during a gruelling tour of duty in the region. One of the anglers, Ansel Jones, an engaging petty officer from Mississippi, is telling me about life at sea and the rococo port calls he and his shipmates have enjoyed. 'We are sailors,' says Jones, with a knowing glint. 'We fit the stereotype.'

As we talk, the rod that Jones has rigged off the stern suddenly clatters into life, dancing staccato on the deck. He dives over to grab it and start trying to reel his catch in. A crowd gathers, already sensing the possibility of an epic fight. Someone hands Jones a special fisherman's harness, into which he snaps the rod for greater purchase and stability. 'I have been waiting for you for two months now,' he calls out in the rough direction of his quarry. 'Come on, you son of a bitch.'

After dragging on a cigarette and gulping down some iced water, Jones settles himself for a long contest. He reels in periodically, trying not to give too much back when the fish resists. Then we see bubbles, followed shortly afterwards by an iridescent flash of silver. Jones manoeuvres his tiring target to the starboard rail, away from the propellers. Suddenly, it appears fully at the surface, prompting a collective gasp of admiration. It's a handsome yellowfin tuna, surely between three and four feet long.

Jones, on the verge of victory, begins a final effort to draw the fish in. It is at the surface almost constantly now, making brief, abortive darts back below. Jones bends his back once more, starting to rip the tuna from the surface of water and towards the deck above. Another crew member is dangling a grappling hook over the side, ready to pull the catch in.

Then, calamitously – at the very moment of success – the line snaps. The crowd groans. The tuna, reprieved at the last, jerks away into the depths. A half-hour struggle has ended in defeat. This succulent fruit of the Gulf of Guinea, which looked ready to be plucked by these American adventurers, is instead gone for good.

Jones takes the frustration well, grimacing as he receives consolatory pats on the back. The problem, he says, is that the ship is too large and cumbersome to land the catch. He needed to be in a smaller boat, where he wouldn't have had to haul such a great weight over the high sides. Flawed logistics and execution meant this West African prize got away from the US military. 'That was a big yellowfin tuna,' he says. 'It would have fed all of us.'

Jones's private Gulf of Guinea battle is part of a much bigger struggle involving the USA at one of the world's greatest and most coveted new oil frontiers. The *Dallas* – a US Coast Guard ship whose more natural milieu is Caribbean drugs busts – is patrolling far from its South Carolina home as part of an American-led initiative called the Africa Partnership Station. Under this, US vessels are maintaining a 'persistent presence' in the seas around the domino run of countries that curve from Senegal on the western edge of the Gulf of Guinea to Angola in the south. US officials say their programme is not just about the oil, although there's too much of it at stake here for them to deny it's an important factor.

All around West Africa a great game is taking place between the USA and emerging economic powers, the result of which will help define the future of big oil and the world it drives. The fulcrum of the contest – as for so long – is Nigeria. Angola may be close to claiming the crown of top producer in the Gulf of Guinea, but Nigeria's reserves are estimated to be four times greater. In an era during which – until recently – oil prices seemed to soar almost daily to historic highs, it pays to stake your claim in the region's oil giant.

As the USA, China, India, Russia and others sniffed around Nigeria, I found it hard to avoid a sense that a disturbingly familiar history was being played out again. The last frenzy of international economic interest in Nigeria had led to the disastrous lending spree that had triggered the financial crisis of the early 1980s. Further back still, I thought of the damaging political power-plays of the Royal Niger Company and its French rivals. Now, as then, the ambition of outsiders seemed straightforward and alarming: they needed to grab their share.

The latest West African mêlée also seemed to me to have at its heart a contradiction that dated back to the birth of Nigeria and had grown increasingly damaging. The established and emerging powers of the world were flocking to take resources from a country that was inherently highly unstable. Many commentators have pointed to the theoretical attractions of Nigeria and its surrounding states as a bulwark against oil-supply problems in the conflict-racked Middle East. Yet when I arrive on board the *Dallas*, a quarter of Nigeria's production is shut down for security reasons. God help us, I thought, if the Niger Delta is the world's insurance against meltdown in the Arab oil states.

My rendezvous point with the *Dallas* is in São Tomé, home of the forest *agua petróleo* and a short hop from the Niger Delta coast. At the main port in the nation's capital, also called São Tomé, US officials are preparing for the *Dallas* visit. Their first job is to receive a small boat from the ship loaded with drinks for a reception being hosted by the USA that night. After it arrives – and the crates of wine and Bud Light beer are brought ashore – there's a small technical hitch, when one of the military officers accidentally drops his car keys into the water. The mighty American forces are saved by a São Tomé officer, who strips off, dons flippers and snorkel and retrieves the lost item from the dark waters.

At the reception, which takes place in the open-air courtyard of the unadorned national parliament building, the islands' notable citizens are served drinks with napkins depicting the US flag. Across the bay, the red lights of the giant Voice of America radio transmission station blink in and out of sight like a nocturnal predator's eyes. Officials at the event hand out sweets and neatly tied copies of the US Declaration of Independence from wicker baskets bought that day from a local trader (at least one item was locally sourced, a US officer jokes to me). A brass band plays 'The Star-Spangled Banner' on ageing instruments, before the cutting of a cake, the icing of which wishes the USA a 'happy 232nd anniversary'.

One of the US officers visiting São Tomé for the occasion is Lieutenant-Colonel Rene Dechaine, a defence attaché based in neighbouring Gabon. Dechaine, an old hand who has spent time in Korea, Brazil and the UK, argues to me that the US attention to Nigeria and the other Gulf of Guinea countries can do good. He rebuts my suggestion that Washington's wanderings in West Africa could be helping to reinforce autocratic governments in a notoriously repressive region. There may be a tension in deciding who to do business with, he says – 'there always is' – but the US programme is about helping make local militaries better servants of democracy, rather than sharper instruments of oppression. Besides, he argues, the appetite of countries for US help is growing. 'Africa is changing,' he says, in a curious echo of the arguments used by supporters of Chinese expansion on the continent. 'They are not looking to the former European masters any more.'

The Africa Partnership Station programme is an expanding mission that is being overseen by the US navy's Sixth Fleet from its base in Naples, Italy. The navy says the initiative is a wide-ranging security effort, aimed at curbing illegal fishing, people-trafficking and drug-smuggling, and many other crimes. The venture's ships have already taken on board scores of local military officers for training and have conducted joint patrols with coastguards from countries they have visited. In separate telephone interviews before I arrived in São Tomé, the navy's Rear Admiral Anthony Kurta and Commodore John Nowell insisted the initiative was neither 'all about the oil' nor solely a US project. Yet, while it is true that officers from other countries have been involved, there is no doubt that this is another international coalition dominated by the USA.

The day after the US drinks reception, I return to the quayside to meet a small boat dispatched from the *Dallas* at its anchor spot offshore. The ship is 378 feet long, about two-thirds the size of a modern destroyer. Its description as a 'coastguard cutter' may sound innocuous, but this is a US warship with a 76mm main gun battery and other tarpaulin-covered weapons scattered around the deck. It is, as bo'sun Jay Bealer puts it to me later, a ship that 'could probably take on most navies in small countries in the world'.

On board, Bob Hendrickson, the second-in-command, tells me I'm allowed to move freely and talk to whomever I want. If I stumble into any secret areas, he adds drily, 'we will just shoot you'. He's as good as his word about the access, although I never get to test the consequences of a mis-step.

Hendrickson, a thoughtful, muscular man seemingly as broad as he is tall, sees the *Dallas*'s mission as something more than the mere diplomacy on display at the previous night's party. He quotes Thomas Friedman, the US journalist perhaps best known for his 1999 'Golden Arches theory' – that two countries with branches of McDonald's in them never go to war (a tongue-in-cheek surmise proved wrong by, among other conflicts, the NATO bombing of Serbia and the fighting between Russia and Georgia). 'What it boils down to is,' Hendrickson says, 'if you don't visit the bad neighbourhoods in the world, the bad neighbourhoods in the world will visit you.' The Gulf of Guinea may not quite be a bad neighbourhood, he adds in a quick switch of metaphors, but its poverty and threadbare security forces make it 'a broken window. And we want to help them fix it.'

The *Dallas*'s other guests for this trip are a couple of members of the São Tomé coastguard and two lieutenants from the Equatorial Guinea navy. The Equatoguineans – Placido Nzang and Pablo Esono – are an interesting choice, given the country's dismal human rights record. President Obiang Nguema Mbasago and his clan have ruled since 1979, amid heavy criticism over election-rigging, corruption and human rights abuses. I'd first visited the country for the 2002 presidential elections, at which Obiang was re-elected with 97.1 per cent of the vote. I'd returned a couple of years later for the trial of some of the alleged mercenaries accused of trying to stage a coup, a hearing that Amnesty International condemned over allegations of torture and many other serious failings. During my visit to the *Dallas*, I hear that Simon Mann, the former SAS officer accused of masterminding the coup, has been jailed for 34 years, in another trial rendered almost meaningless by the system under which it took place.

For the next two days, the *Dallas* patrols the triangle of open sea between Nigeria, Equatorial Guinea and São Tomé. These are among the most hazardous waters in the world, according to the International Maritime Bureau, which ranked Nigeria as second only to Somalia in terms of acts of piracy during the first three quarters of 2008. The biggest business is the theft of oil, particularly from Nigeria. It's estimated that many hundreds of millions – or even billions – of dollars of oil have been taken over the past eight years from the creeks of the Niger Delta, loaded into tankers off the Nigerian coast and taken to refineries in the region and beyond for onward distribution in the black market. The trade funds the armed militants in the Niger Delta, who are increasingly taking oil workers hostage and attacking installations.

The great nexus between African piracy, lawless states and oil was highlighted spectacularly in December 2008, when a gang operating off the coast of Somalia seized the Saudi Arabian tanker *Sirius Star*. The ship was laden with 2m barrels, about equivalent to Nigeria's daily production. The hijack soon attracted a swarm of private security experts, many of whom had first-hand experience of Nigeria's distinct but parallel problems of criminal gangs operating without interference by local officials. As one kidnap and ransom specialist put it, in a remark that might apply equally to Somalia or Nigeria, 'The authorities know who these guys are. They know them right down to the individuals.'

Just before I'd arrived in São Tomé, Nigerian militants had provided their own reminder of the kind of conflicts in which the *Dallas* and other visiting US ships might become involved. The militants had targeted Shell's massive Bonga oilfield 75 miles out to sea, cutting national production by 10 per cent. The attack – the first ever launched so far offshore – sent a tremor of fear through a region in which much of the oil is sunk deep under the ocean. After I leave, the *Dallas* will carry out an exercise simulating the defence of an Equatorial Guinea offshore oil platform operated by Marathon, the US multinational.

Oil theft is one of the problems the US Africa Partnership Station mission explicitly says it is trying to tackle. Washington's call to action has been echoed by Nigeria's president, Umaru Yar'Adua, who has compared the trade to the so-called 'conflict diamond' dealings that have fuelled brutal wars in countries such as Sierra Leone. He found a sympathetic ear in the British prime minister, Gordon Brown, who offered in July 2008 to provide military assistance to help stop the illicit trade. That largesse achieved the unusual feat of dismaying both British army chiefs and human rights campaigners in the Delta. One activist observed how Brown's remarks dangerously upped the military ante in the region, and ignored Nigerian officials' complicity in both oil theft and human rights abuses there. It reminded me of a conversation I'd had with a US embassy spokesperson early in my time in Lagos, when Washington had announced that it was donating patrol vessels to the Nigerian military for use in the Niger Delta. When I'd asked how the USA was sure the craft weren't going to be used to shoot civilians and destroy villages, she had replied that she'd wondered that herself.

Out on the *Dallas*'s deck, looking around at the empty grey swell, I can see why pirates and oil thieves have free rein in these seas. Although we are deep in the heart of the tropics – the Equator runs through São Tomé – this isn't a region of pretty and populous little islands. During my five days on board, we see little evidence of humanity outside the ship, apart from a few trawlers and the odd small fishing boat. At one point, I see Jimmy Tiny, one of the São Tomé coastguard officers, leaning over the starboard rail to be sick into the water. He tells me that it's his first time in such a big boat so far from home.

Early on my second day on the *Dallas*, I receive a call on the public address system to go to the bridge immediately. When I arrive, the room is humming as I've not seen it before. Captain Wagner is plotting a course to intercept a Japanese vessel that is looming rapidly to port from the grey morning sky. It is the first catch of the day of the mission's Gulf of Guinea trawl.

A few moments later, Jay Hollister, chief warrant officer, bursts through the door at the back of the bridge and strides over to the radio. Hollister has been summoned to talk to the other vessel, the *Kaichi Maru*. He is married to a Japanese woman and speaks the language a bit, although he is far from fluent. He starts a halting interrogation, asking in Japanese whether anyone on board speaks English. 'Japanese only,' comes the reply, somewhat paradoxically in English. 'I'm sorry.'

Hollister presses on in Japanese and reports back to his shipmates that the *Kaichi Maru* claims to be a trawler, operating under licence from São Tomé. 'In that case,' retorts Captain Wagner, 'ask him where he has set his fishing gear.' Hollister, still a little shy, says he's going to have trouble asking the question. 'I don't know how to say *fishing* as an adverb,' he explains. 'I know how to say *fish* – but not *fishing*.'

After a little improvization – 'Can you say to him, "Where are your nets?"' suggests one fellow officer – the point is put across successfully. The radio operator on the *Kaichi Maru* – seemingly unfazed by the US warship bearing down on him – explains calmly that the crew are not working because they have already filled the hold. He offers to send the *Dallas* a copy of the vessel's fishing licence, but Captain Wagner tells him not to bother. He waves away Hollister's apologies for his shaky linguistic skills. 'You did more than I did,' he says, with a laugh.

'So,' the captain continues, turning to address the whole bridge. 'Let's go find something else.'

I ask Captain Wagner later what he feels he would have the authority to do if he came across a kidnapping or attack on an oil installation. Lawyers say a complex blend of international agreements and domestic rules governing piracy is causing disagreements about when and how other ships can intervene. Some countries feel that they do not have the power for their ships to act under the 26-year-old United Nations Convention on the Law of the Sea, the framework for international maritime rules. The convention authorizes 'repression of piracy', but only on the high seas rather than in national territorial waters. It also threatens states with financial penalties if they seize ships 'without adequate grounds'.

Captain Wagner acknowledges his powers are limited while operating in the Gulf of Guinea, although he says he thinks he would be able to respond to an incident of piracy. I ask him whether a Nigerian-militant-style assault on an oil platform would qualify as such and justify intervention. He replies carefully that he 'would have to ask the lawyers the definition'. After a moment's further thought, he says, 'If someone was in that position, there would be more of a basis for action than for inaction.' He adds, 'If something happens, we are the only US warship in this part of the world.'

The region that the *Dallas* is patrolling is a key theatre in the quest to satisfy growing world demand for crude. The US Energy Information Administration has forecast that world oil consumption will rise 37 per cent between 2006 and 2030, although that estimate was made before the global financial crisis and the consequent sharp fall in prices. The Administration forecast that West African production would climb 72 per cent over the same period, more than double the rate of expansion of Middle Eastern output. New Gulf of Guinea production – whether in established countries such as Nigeria or virgin nations such as São Tomé – is essential to achieving those growth rates. In August 2008, the countries of the region were providing a shade under 15 per cent of US oil imports, with more than three-fifths of that coming from Nigeria.

The global rush to develop new West African energy projects and expand old ones has brought with it fears of an unregulated free-for-all in an industry already dogged by examples of proven and suspected corruption. That applies most of all to Nigeria, which delivers roughly half its oil to the USA. The most high-profile oil graft case of the past few years has been the US investigation into whether an international construction consortium, including an ex-Halliburton subsidiary, agreed to pay $170m of bribes to win billions of dollars of work on Nigeria's giant Bonny Island gas liquefaction plant. In September 2008, Jack Stanley, a former top executive at the consortium, admitted overseeing an eight-year bribery scheme involving tens of millions of dollars and officials at senior levels of three successive Nigerian administrations, including the internationally acclaimed civilian government of President Olusegun Obasanjo. The gas plant's private sector multinational owners – Shell, Total and Eni – have taken no action to date against the consortium over the allegations, instead giving it more than $1bn of extra work. Halliburton and a former subsidiary, KBR, agreed in February 2009 to pay $579m in penalties to the US authorities over the case, the second-largest fine in history for corruption.

Other US probes have identified wrongdoing in Nigeria by big oil industry names such as Baker Hughes, Vetco Grey and Willbros, the unwitting donor of Asari's headgear. Few people expect these companies to be the last thus embarrassed. It all brings to mind the comment a journalist colleague of mine made to me after discovering that one of the collapsed US energy company Enron's many alleged misdeeds was to fake a sale of power plants mounted on barges off Nigeria's coast. Enron and Nigeria, he noted, seemed two particularly well-matched business partners.

Not many anti-corruption campaigners think the arrival of the world's new economic powers in the Gulf of Guinea will help improve behaviour. Instead, activists talk gloomily of their fears of a 'race to the bottom', as the Western multinationals and their new challengers compete ever more aggressively for business. When Nigeria auctioned off 25 exploration blocks in 2006 – many to companies from Asian countries keen to expand their energy resources – the *Financial Times* reported widespread allegations of 'political favouritism and back-room dealing'. In December 2008, research by Transparency International, the anti-corruption group, said companies from China, India, Russia, Mexico and Brazil were perceived as the worst offenders among leading world economies in terms of paying bribes to win business overseas. Of the five, only Mexico and Brazil have signed up to a key international anti-bribery convention overseen by the Organisation for Economic Co-operation and Development, the Paris-based club of big industrialized and industrializing countries.

As my days on the *Dallas* go by, I see increasingly how its mission reflects both the extent and limits of US influence as it competes in energy-rich West Africa with the nations that threaten its global economic dominance. The ship is solid, but more than 40 years old: it's already had one refit and is due to be replaced by a more modern craft. An engine-room fire just before it arrived in São Tomé left it short of lubricating oil, meaning that it has to cut its speed to as little as five knots until it can restock. The handicap is accentuated by the lack of intelligence on local ship movements from São Tomé's new US-backed radar facility, which is closed at the weekend. The combination of setbacks makes the *Dallas* a crippled giant, at times flailing blindly like Polyphemus in the hope of encountering other vessels.

The onboard training programme for the guest military officers has a similarly desultory quality that seems to stem from both bad planning and bad luck. A lesson in basic navigation is a Babel-like affair, in which each of the teacher's statements in English is translated into Portuguese for the two São Tomé officers and Spanish for Lieutenant Esono. A firearms course has to be scrapped after the São Toméan fusiliers who wanted to come on board are refused entry because they haven't been vetted by the US authorities. Another security training session has to be postponed because one of the São Tomé officers is feeling seasick.

The São Toméan officers do receive a basic training in handcuffing and the use of batons, in neither of which they seem very experienced. The students are given a handout that sets out an escalating scale of aggression levels they can use, culminating in lethal force. The trainers say they always have to be flexible, because of the different political and cultural traditions from which their subjects come. In the Cape Verde islands, the *Dallas*'s last stop, the officers said they never needed to handcuff people. Other visitors are less reticent. One trainer recalls a member of the Albanian security forces who told him that, if suspects didn't listen, he would 'pull out a gun and put it in their face'. And if they still didn't listen he'd pull the trigger.

Perhaps the *Dallas*'s greatest problem is the shortage of linguistic skills on board, which makes it tough to deal with the cosmopolitan range of other vessels plying the Gulf of Guinea's waters. This is a big theme of what I fancied was my mildly samizdat choice of book to read on board: *Legacy of Ashes,* Tim Weiner's magisterial history of the Central Intelligence Agency. One of Weiner's laments is the lack of linguistically gifted and culturally savvy CIA officers able to penetrate the Soviet Union and other countries that were or are largely a mystery to the West. A few members of the *Dallas* company can speak major Western European languages, but Captain Wagner admits it's not enough. 'One of the biggest shortcomings of Americans is languages. We don't speak enough,' he says, although he insists the situation is improving.

The tangible uncertainty among the US officers over how to handle this new and unusual mission seems to be reflected in the ranks. Very few of those on board the *Dallas* – which is based in Charleston, South Carolina – have been hand-picked for the venture. The crew is a mix of career coastguards and young people who say they plan to do a few years in the service and then move on to something else. There's a wannabe chef, a young man who plans to set up a bar, and an engaging if dreamy writer who reports with delight Commander Hendrickson's description of him as the ship's 'warrior poet'. When I tell one crew member that I was surprised to hear a US coastguard ship was headed for the oilfields of West Africa, her reply is disarming and enlightening. 'So was I.'

All the crew members I meet aboard are polite to me and many are friendly and open, despite my endless questions about them and what they do. It helps that I'm a rare new face on a cramped four-and-half-month voyage in which one person says the only way to cope with seeing the same people all day every day is to 'just think of them as those strange cousins you never really cared for'. One seaman who is under 21 – the drinking age in some US states – is excited to be in a part of the world where the qualification for buying alcohol is 'if you are tall enough to reach the bar'. Another says he's on board because he wanted to travel but didn't really like killing people. 'So I didn't want to join the army,' he says.

Although the ship's company have all been told that details of the mission, politics and personal opinions are off limits in their conversations with me, our talks inevitably stray. Some of the crew start speaking vaguely about the point of the *Dallas* visit – 'to get our face out here, show Africa we care' – before veering off in more interesting directions. A number are worried about the international image of the USA post-Iraq. One sees the aim of the West Africa visit as 'trying to make friends, make democracy – and trying not to make us the bad guy that other countries think we are'.

A number of those on board are contemplative about the USA's place in a world in which its hegemony is fading. Some seem caught in the bind that seemed to grip the George W. Bush administration during its dying days, veering from swaggering confidence on the one hand to manifest insecurity on the other. On the mess deck one night, a young seaman proselytizes to me about the United States' 'responsibility to help lesser nations'. But then, in almost the same breath, he speaks of the urgent need to convince the world that US citizens are not 'arrogant assholes making war everywhere'. 'I don't like living in the USA,' he continues. 'Everybody's arrogant. Everybody thinks they are better than anybody else. Other countries don't think that. They are not conceited.'

Captain Wagner himself is diplomatic, playing up the ambassadorial side of the *Dallas*'s West Africa trip and for the most part steering away from its more contentious aspects. He is undoubtedly the man for the job: he's a motivator who is constantly on the move around the ship, his head habitually inclined because he is a tall man in a low-roofed vessel. He has a sticker on his door – put there, he assures me, by someone else – that describes him as 'Captain Awesome'. Next to his computer is a sign: 'There are three kinds of people. Those who make things happen, those who watch things happen and those who wonder what happened.'

The captain says – with a laugh – that he hopes to use the Gulf of Guinea mission to 'dispel a few myths' about the USA and 'hopefully not reinforce anything that's bad'. Like his number two, Commander Hendrickson, he invokes a contentious – some would say glib – aphorism popularized by Thomas Friedman. 'The world is flat these days,' the captain says, meaning that even powerful nations have to build trust with other countries. On a personal level, his overseas missions have opened his eyes to all sorts of things, not least the USA's uniquely large appetite – at least until the international financial crisis took hold – for gas-guzzling vehicles and other consumerist trappings. 'Sometimes I wish I could bring my kids over here and show them a couple of things,' he says. 'You don't have to have electronic games to have a good time.'

Yet, for all the warm and fuzzy aspects of the *Dallas* mission as articulated by captain and crew, its harder strategic edge is never far from view. The longer I stay on board, the more I realize how much of its unthreatening appearance is artifice rather than accident. The captain is one of several people to point out that the vessel's paintwork – which is white instead of navy grey – makes people feel a bit more at ease than if there were a US cruiser anchored off the coast. That sleight of hand allows it go to places the navy cannot reach. (The ploy isn't foolproof, as the *Dallas* itself found during the Russia–Georgia conflict in late August 2008, when it had to hastily reroute to avoid a face-off with Moscow at Georgia's Russian-controlled port of Poti.)

The latent realpolitik surfaces from time to time in conversations with US officers, who seem resigned that, whatever they say to the contrary, many people will still believe their interest in the Gulf of Guinea is all about oil. Lieutenant Frank Florio, who accompanied Commodore John Nowell on an earlier Africa Partnership Station trip to the Gulf of Guinea in the USS *Fort McHenry*, including a stop in Nigeria, says, 'I don't think we are ever going to beat that. Everywhere the commodore goes, he gets asked about oil.'

The limping *Dallas* finally stops to restock on lubricants at Port-Gentil, oil capital of Gabon. The ship's senior officers take the opportunity to host a lunch there with top members of the local navy. Gabon is another oil-rich Gulf of Guinea state that is not noted for cleaving too closely to the democratic ideals of the US founding fathers. When Fidel Castro stood down from the Cuban presidency in 2008, Gabon's president, Omar Bongo, became the world's longest-ruling executive leader. He'd held power since 1967, overseeing a state in which corruption has flourished and the now-declining oil reserves have been squandered with little to show in terms of long-term infrastructural development.

The president prospered for many years owing to his close relations with France, the former colonial power, and the oil company Elf, a number of whose senior executives were jailed in 2003 for their involvement in Gabon-related corruption. Britain's *Guardian* newspaper has described the case as 'probably the biggest political and corporate sleaze scandal to hit a Western democracy since the Second World War'. When I tell one of the *Dallas*'s officers about Bongo's history and longevity in Gabon, I receive a nonplussed response: 'Do they like him there?'

As we pull out of Port-Gentil, trying to avoid the logs from a local sawmill that at times float perilously close, I chat to Lieutenant Frank Montalvo, the ship's third in command. We're leaning on the rail on the deck to the port side of the bridge, in front of a gyrocompass and a padded white chair specially reserved for the captain. Montalvo, a chunkily built Texan often seen working out on the *Dallas*'s rowing machine, has a thoughtful take on life with the military at sea. He's particularly interested in the sensibility of the younger Americans in the crew, and how they have little awareness of conflict beyond Iraq or – at most – the first Gulf War.

I tell Montalvo that this trip has also made me think about the history of conflict. I say I see parallels between the chaos of Nigeria – and its fellow Gulf of Guinea oil-producing states – and the perfect storm of wildness and corruption that Daniel Yergin's classic 1990 book *The Prize: The Epic Quest for Oil, Money and Power* describes as characterizing the birth of the US crude industry in the nineteenth century. Montalvo agrees that people tend to forget these things, looking instead at Africa's leading oil producers as uniquely awful and saying, 'I can't believe this is going on.' He says, 'Well, you know, humanity has a way of repeating certain patterns.'

Montalvo says he's seen a headline in a newspaper back home that claims Africom – the United States' African military command established in 2007 – is about oil. Of course that's a part of it, he admits: the USA needs crude, which has 'a way of maintaining stability in the greater planet'. But this is a mission where 'corporate interests and human interests have a lot more in common than people give them credit for', he insists. The initiative has 'tensions' – Lieutenant-Colonel Dechaine's word again – in areas such as deciding whether it's right or not to work with certain countries, but the USA is determined to be neither censorious nor patronizing. 'It's very important for us to keep in mind that we have had these problems too,' Montalvo says, as Port-Gentil and its huge oil refinery slip behind us. 'We shouldn't be taking a high and mighty tone, saying, "OK, here you go, let's explain it to you, let's write it down in crayons."'

On my last night aboard the *Dallas*, Captain Wagner invites me to dinner at his table. Over shrimp gumbo, eaten under the watchful eye of a portrait of Alexander Dallas, the nineteenth-century US Treasury secretary for whom the ship is named, the captain talks about how the US navy is trying to 'create stability in a region before there are problems'. If it can persuade people to work together over 'small things' such as fishing and human trafficking, he says, maybe it can get them to 'work together on other things to prevent larger escalating conflict'. 'It's one of those things where you may not see the thing tomorrow,' he says. 'But I think you will see the return a couple of years down the road.'

The captain wants to know more about my experiences in Nigeria and particularly my encounters with militants in the Niger Delta. I say that his ship could have an interesting time scouring the Nigerian coast, where the vessels that distribute stolen oil around the world are said to gather. The captain seems taken by the idea and a little regretful that – along with Angola – Nigeria is an important country the mission hasn't been able to visit this time. He could conduct a pretty good survey of the coastline in a couple of days, he says. That would be fascinating, I tell him: the question is whether the corrupt domestic and expatriate interests that have fed so richly off the smuggling for so long would ever let him do it.

The captain is upbeat about the *Dallas*'s West Africa trip, even though he acknowledges the São Tomé visit was a 'major disappointment' because of the technical problems and the difficulties arranging meetings there. These US African missions are, he says, a way of 'mending fences' and creating opportunities for further joint activities. The USA is not ending wars – or, he might have added, starting them – but it is helping build important relationships for the future. He tells me about his uncle, who was on board the USS *Missouri*, the battleship on which Japan surrendered in 1945. 'That closed one chapter in history and opened another,' he says. 'This is kind of the same thing.'

In the officers' mess after breakfast the next morning, my last on board, the conversation turns to the significance of the *Dallas*'s earlier stop in the Cape Verde Islands. I soon realize that this was another visit to a small, remote and apparently obscure Atlantic archipelago that had more resonance than I'd first detected. Cape Verde is symbolic because it was once used as a major transatlantic stopping-off point, at first for British slave ships and then for coal-powered vessels in the nineteenth century. Some of the coal came from Cardiff, but, as ships switched from coal to oil, Britain lost interest in Cape Verde as a refuelling stop. In other words, an important West African energy resource, once firmly within the West's sphere of influence, these days has a more ambivalent status.

Now, say many who visit Cape Verde, the more contemporary story is of the rise of China. One researcher has dubbed it 'the Dragon's African paradise'. Years of immigration by small-scale Chinese merchants is being buttressed by large investments in sectors from casinos to cement. The Chinese government has agreed to fund the construction of the national parliament, stadium and library, plus two dams, a fleet of vessels to link the islands and a medical team in the main hospital. It's an investment that, as one member of the *Dallas* company acknowledges, the USA is unable to match.

'It can't compete with the Chinese shipping over and building a hospital,' he says. 'Africa is the last place up for grabs. That's why you find the Chinese hitting it huge.'

In Nigeria, China has won oil exploration deals in part because it has pledged big infrastructural investments in projects such as gas pipelines and refineries. It has also offered a $2.5bn loan to finance a railway. All have been slow to materialize, raising the question of whether they will be stymied by the international financial crisis and recession. In December 2008, the *Financial Times* reported that Dangote Group, a sprawling Nigerian

conglomerate, had postponed plans to buy $3.3bn of cement plant build-
ing and materials from Sinoma International Engineering, a Chinese
contractor.

The idea of a new international scramble for Africa – and in particular for its energy reserves – is hardly original. But what's striking
up close is how intense and amoral the process is. Both the USA
and China are wooing countries in the region, including Nigeria,
that have poor records on holding credible elections and respecting
human rights. In the most egregious cases, the powers of both East
and West are snug with nations that, in any principled approach to
foreign affairs, would surely be international pariahs to match Robert
Mugabe's Zimbabwe. After the *Dallas*'s stop in repressive Equatorial
Guinea, I receive an e-mail from someone on the ship telling me of
his shock at seeing how the Chinese are 'taking over' the country with
a blitz of construction work.

All this activity looked to me like a huge international bet on the
Gulf of Guinea countries – and Nigeria in particular – as reliable
sources of fossil fuels. It seemed a cynical policy that was even more
questionable because it was dubious when judged even on its own
pragmatic terms. Missions such as the Africa Partnership Station
would do nothing to dispel feelings of injustice on the ground in
the Delta; if anything, as Britain's colonial experience showed, an
ostentatious military presence was likely to increase them. The lesson
of the region was that companies were finding it harder and harder
to take the oil mainly because of the very conditions created by half a
century of the industry's operations there. Unless that was addressed,
many Nigerians found it hard to see much of a future for the country
as a viable state, from the point of view of either its inhabitants or
foreign investors.

Not surprisingly, this isn't a message that has emerged too often
during the past five years, when the (mostly) soaring oil price gave
the most powerful countries of the world a strong interest in improving their relationships with Nigeria. From about 2004 onwards, as
the international crude market began its long upward climb, the
Western diplomatic narrative of the nation switched from a story of
near despair and exasperation sometimes bordering on contempt to

a tale of reform and increasing prosperity. Foreign companies once again flocked to do business in a country flush with wealth from the combination of debt relief and high oil prices. In a valedictory interview in 2007, President Olusegun Obasanjo celebrated the emergence during the last part of his eight years in office of a class of super-wealthy, politically connected tycoons, who were increasingly compared with Russia's post-Soviet oligarchs. 'Why can't we have a Nigerian among the three richest persons in the world in the next ten years?' the departing leader asked, seemingly not afraid to echo the hubris of past Nigerian oil booms that turned to dust.

As I prepare to leave the *Dallas*, I take a final look around haunts that have already become too familiar after five days on board. On the deck overlooking the rail where Petty Officer Jones fought in vain, I spot the Equatoguinean officers sitting next to the rowing machine, heads down and looking a little bored. I ask whether they've learned a lot that will be of practical use for their future work. Lieutenant Nzang looks up and tells me, with a scornful brevity I find rather sinister, that they haven't learned all these skills just to '*charlar*', or chat.

After I finish with the officers, I walk round the corner and lean on the side of the ship to note down what was said. As I'm doing so, Clifton Chang, a young crew member, pops his head round the door of the nearby weights room. We start to talk and he asks me about Nigeria, as I'd told him I used to live there. I speak about the problems surrounding the country's oil industry.

When I've finished, Chang asks me, 'Do we have any connection with oil?'

'You tell me,' I reply with a laugh.

Chang says nothing. The thought is still hanging in the air a few moments later, as he turns back to the metal gym frame to resume his pull-ups.

8

THINGS ARE LOOKING UP

As battle headquarters go, Shell's hulking complex on the south bank of London's River Thames is pretty imposing. Nestled near Waterloo station and just downriver from the Houses of Parliament, the office is a short step from the centre of Establishment London. It is here – as well as in The Hague – that the company's chiefs over the decades have overseen its expansion into one of the handful of so-called 'supermajors' at the head of the world oil industry, notching profits of $31.4bn in 2008, a record for a European company. One of the main countries underpinning that growth has been Nigeria.

Shell's Waterloo base is also the most calm and secure company waterside facility that I've seen since I started visiting the Niger Delta. On the way there, relishing a rare fine English spring morning, I'd passed a busker under nearby Blackfriars Bridge tapping out the Beatles' 'Yesterday' on his xylophone. It certainly was the kind of day when the Nigerian troubles I had come to talk about seemed remote.

Inside Shell HQ, Basil Omiyi, the company's managing director for Nigeria, is waiting for me in a meeting room on one of the upper floors. He's bathed in sunlight and seems relaxed enough, but the mood changes as soon as I ask him how things are. 'Difficult, as you know,' he replies. 'Difficult. But things are looking up. We have to be optimistic about the state of the country. There is always this huge negotiating space.'

It's the remark of a man who knows – sitting talking to me in 2008 – that he's at the heart of troubles so deep he can't pretend they don't exist. Shell, more than any other multinational, has come to represent the past, present and future of oil in Nigeria. Shell is the pioneer, the biggest of the companies operating in the Delta and – thanks to the hanging of Ken Saro-Wiwa and eight other activists in 1995 – the most notorious. Yet, for all the omnipotence its more conspiratorial opponents ascribe to it, its fortunes and standing have ebbed during an era of armed militias, pollution, corruption and sabotage. Big oil is under pressure to change on these fronts in other places, too, but there aren't too many countries where there is such a perfect storm of both troubles and criticism of the industry for its role in them.

Omiyi's low-key assessment also carries with it a sense of the souring of the reform story that the international business and political communities had been telling about Nigeria over the preceding few years. Under Olusegun Obasanjo, president until 2007, and during the early days of his replacement, Umaru Yar'Adua, Nigeria's international image had undergone an extraordinary change. Dismissed for years by diplomats as a basket-case – or, in one memorable metaphor, an off-course supertanker that would take at least a generation to turn around – it had suddenly switched to being Africa's great hope. Britain tripled aid and stepped up the rhetoric of recovery so much that even one Whitehall official at the heart of it all admitted privately that there was 'a bit of a disconnect between this very positive message that goes out and the feeling on the ground'.

In one breathless article in Britain's *Independent* newspaper, Ngozi Okonjo-Iweala – a finance minister who had made some important but limited changes to how the country's oil wealth was managed – was elevated to the status of a woman who could save the whole of Africa. It seemed as if the world's view of Nigeria had switched from irredeemable disaster to leader of a renaissance on the continent, without stopping to consider that it might be neither one nor the other. The miraculous turnaround became the new orthodoxy.

Omiyi's ambitions are more modest: to be the positive face of a business that is supposed to pump more than 1m barrels of oil a day, but has been plagued for years by violent production disruptions. Age seems to be treating him kindly, as does the smile that takes years off his already youthful features. His shirt is crisp and white and his dark suit snug on his full figure. Altogether, there is something old-school about him, his formal manners and mild chubbiness hinting at prosperity and a certain social standing. He likes to talk, too: during the two hours of our meeting, he gives long, thoughtful answers that don't always resist sententiousness.

Nigeria is a key sphere of Shell's operations. It was the company's top country for crude oil production in 2007, accounting for 16 per cent of worldwide output. Nigeria is also a crucial source of new oil and gas. When the company became embroiled in scandal in 2004 for overstating its reserves worldwide, it turned out that Nigeria accounted for a third of the incorrect claim. Shell has a much higher public profile in the country than companies such as ExxonMobil, which mostly operates offshore rather than in the thick of the creeks. Shell's Delta operations include 90 oilfields, 1,000 wells and more than 3,750 miles of flowlines and pipelines – almost long enough to drop down to the centre of the earth. Not for nothing are parts of the Niger Delta sometimes referred to, jokingly, as the 'Republic of Shell'.

Omiyi joined Shell in 1970, immediately after the despair of Nigeria's three-year civil war. His boyhood dream had been to become a commercial pilot, but the company came recruiting just as he was finishing his engineering degree. He says Shell's ethos resonated with the 'very strong religious values' of his Catholic upbringing. He says he relished the business's 'integrity, and respect for people, honesty'. It wasn't long before he was overseeing a drilling programme worth millions of dollars. He was 'fully in charge', unlike Nigerian peers at other oil companies, who always had 'someone looking over their shoulder from Milan' – or wherever head office was based. Now he is the first Nigerian to head Shell's operations in the country, after years in which the company – like its rivals – was criticized for giving the top jobs only to expatriates.

Omiyi claims Shell's roots in Nigeria are so deep that it is widely seen now as a Nigerian business. The comment sounds self-serving, especially given the attacks aimed at the company over the years, but there is a glimmer of truth in it. The business is a large and profitable enterprise in a country of very few, meaning that people seeking jobs and other benefits often gravitate towards it. 'Even the Nigerian populace treat Shell as a citizen of Nigeria,' Omiyi says. 'I think because we are big and old they expect a lot more from us, at times way beyond what business can do.'

If Omiyi makes the relationship between Shell and Nigerians sound like an unlikely – if unbalanced – love affair, then it is one that has become increasingly fractious. The atmosphere in Delta communities has, he acknowledges, deteriorated sharply since the days when he first started working for the company. Then, its officials would drive around freely, even at night. Now they are under curfew, forced to return to their luxurious compounds as soon as darkness falls. 'I would say the communities were quite friendly,' Omiyi says of the past; 'a lot friendlier than they are today.'

In the 1970s, when oil prices were booming in the years after the Arab oil embargo, money had flowed into oil producing communities via both the companies and government. Official spending was lavish and included grants for scholarships for overseas study, Omiyi says, clasping his hands together as if to wring them in regret at good times past. The country had seemed destined to become an economic world power on the back of its oil wealth. But, as Nigerians now know too well, money was borrowed so recklessly that the fall in the oil price in the 1980s led to economic collapse.

The problem for Nigeria – and, ultimately, its oil multinationals – was that living on the patronage of crude meant dying by it, too. The plunge in the oil price undermined all spending that depended on its remaining high. Scholarships, hospitals and roads were no longer provided by the government, so industry – which, unlike the authorities, still had money – moved into the gap. The companies were 'gradually being drawn step by step into providing social service', Omiyi says, pushing his chair back and folding his arms, as if in disgust. 'To be frank with you, the communities were not prepared . . . to own this infrastructure,' he continues. 'So very soon it runs into disrepair and causes a lot of friction between us – because while we are building more we have to keep the old ones running.'

Critics of the oil industry have a rather different take on its tactics:

they say it entered into a Faustian pact with Delta communities that has proved hugely damaging to Nigeria and, in the long run, to the companies themselves. Essentially, the argument goes, the companies bought off villages – sometimes with projects, sometimes with direct cash payments – to pacify them for as long as the oil flowed. It was an easy, *ad hoc* strategy to pursue in a country destabilized by the chaos of military coups and corruption. Omiyi himself has some choice words about the levels of graft in those days, criticizing a senior official of the time in a remark he immediately seeks to withdraw. ('I have got to shake his hand when I see him,' he explains, a little sheepishly.)

The growing community disillusionment with the oil industry finally burst into the world's view during the Ogoni crisis, which led to heavy criticism of Shell. Omiyi denies Shell has responsibility for the greatest crime widely laid at its door: the execution of the 'Saro-Wiwa nine'. As he talks, he raises his finger in the air as if scolding a child who simply won't get the right idea. The Ogoni struggle was not about Shell but about autonomy, he says, pointing to a bill of rights signed in 1990 by Ogoni leaders including Ken Saro-Wiwa. The document does call for the Ogoni people to be allowed to manage their own affairs in a 'distinct and separate unit' within Nigeria, but Omiyi does not mention the anger it expresses over the results of Shell's decades of oil production. The bill of rights describes the area as an 'ecological disaster', and one where people have no water, no electricity and no jobs. 'The Ogoni people have received NOTHING,' it says.

The Ijaw uprising that followed Ken Saro-Wiwa's hanging emphasised how oil production in Nigeria was becoming an ever more lethal business, with the role of the oil companies and their links with the state called increasingly into question. All the multinationals operate joint ventures with the government, in which they have only minority stakes. Many Nigerians see the industry as too closely connected to the executive and to national security forces that have a popular reputation for brutality and impunity. According to official figures, the national police killed more than 3,000 armed robbery 'suspects' in 2003 alone as part of a crackdown known as 'Operation Fire for Fire', a state-sponsored death toll equivalent to that of the terrorist outrage of 11 September 2001.

One illustration of how the boundaries are blurred between the state and the oil companies' operations is a curious, Orwellian-sounding organization called the Supernumerary Police. These police – drawn from the national force – are a crucial prop to the industry whose installations they guard. According to the multinationals, they are employed by the Nigerian state but are allocated to work for the companies in jobs such as driving, under an existing law that provides for such an arrangement. Shell says such officers do not carry guns, although some activists dispute this.

It's hard not to regard these officers – who go by the evocative abbreviation of 'Spy police' – as anything but a state-backed security corps for the oil industry. Many Spy regulars like working for the multinationals, as, unlike the national police, the companies pay salaries on time and offer other valuable benefits such as housing and healthcare. On one visit to Shell's Port Harcourt office complex, I saw a sticker with the silhouette of an officer and the slogan 'Proud to be a SPY-policeman'. 'Be practical, be obedient, be loyal . . . be courteous, be efficient,' ran the message, which was printed alongside the company's distinctive logo.

Shell has also had some intriguing security connections with the very militant groups who stop its production and kidnap its workers. In 2006, the *Financial Times* revealed that the multinational had subcontracted work to companies run by activists involved in a 2003 uprising that shut down 40 per cent of the country's oil output. Shell used the two businesses – Shad-Ro Services and Integrate [*sic*] Production System Surveillance (IPSS) – for waste disposal and pipeline security work. The companies' bosses – Shadrack Otuaro of Shad-Ro and Messio German of IPSS – both had links to the Movement for the Emancipation of the Niger Delta (MEND), the main umbrella rebel movement in the region. Shell said at the time that it negotiated all agreements in good faith and continued to 'actively seek the involvement of local communities in the contracting process'.

Many activists see these kinds of arm's-length relationships as an attempt by the multinationals to ensure strong protection of their facilities and neutralize potential antagonists while retaining the ability to remain disconnected from any human rights abuses. In other words, it is a kind of sophistry that allows big business to reconcile the dirty, sometimes deadly, business of extracting oil in Nigeria with the need to be able to demonstrate 'good corporate citizenship' to Western audiences. 'When there is a problem [with Spy police] they will distance themselves,' says one veteran activist. 'They will say, "That's a police matter." So on the question of trying to punish somebody, they will stand aloof.'

Omiyi is certainly at pains to dissociate Shell from the violence that has flared around its operations and those of the wider industry. He says the company has always told the government of its commitment to human rights, and has worked with Pax Christi, the international Catholic peace organization, to establish standards of behaviour for the Nigerian security forces. Officers involved in wrongdoing are disciplined and, if necessary, removed. Far from being a promoter of abuses, Shell helps to curb them, Omiyi says. 'Today, if you still talk to Nigerian security forces, they say, "Ah, you don't go to Shell because they will ask you to fold your arms behind your back and be shot at and do nothing."'

His comments are far more conciliatory than those made privately by many oil workers, some of whom call unashamedly for state-sponsored repression. I'd had a particularly candid conversation in a Port Harcourt bar the previous year with three engineers, one of whom was with Shell. All were veterans who had seen plenty of trouble in the Delta and, like half-crushed cockroaches, just kept on coming back from it. As we sipped from big brown bottles of Gulder beer, the Shell man speculated that a military coup might make the place a little less lawless, while one of his colleagues called for the country to be given a dose of benevolent dictatorship. The third member of the trio had the most eye-opening prescription of all for this nation in which he was a mere guest worker. 'You need a government of businessmen,' he told me. 'And,' he added, lifting his right hand and squeezing it slowly and menacingly into a fist as he spoke, 'a mercenary army.'

Omiyi, by contrast, will go no further than the bland observation that the security situation in the Delta has grown much trickier for Shell over the past five years. It 'really became very difficult' for the company, he says, once the government sent troops to the region in 2003. This deployment – which happened shortly after he became Shell's production director for Nigeria – was a response to the latest in an ever more serious succession of violent flashpoints in the region. The dispute – between the Ijaw and the Itsekiri, another ethnic group, over the alleged gerrymandering of some local government boundaries – led to scores of deaths and the temporary shutdown of one-third of the country's oil production.

The problem for Shell and the other oil companies, according to Omiyi, was that the soldiers inevitably came to them to seek logistical support. The military were operating in creeks where there was 'no place to get water to drink' and 'no place to get connected to power', except at oil installations. So Shell helped them. 'We were the only ones with any semblance of life support systems out there,' Omiyi says. 'You couldn't tell a nation that is trying to maintain peace between its people that you will not allow them to have access to your borehole for water.'

He gives short shrift to the argument that perhaps Shell should have told the government exactly that, given the military's record. Nobody wanted to stop oil production, he says, adding that the company's workers had a right to be protected. The industry did not invite the soldiers to the Delta, but if the military were ever to pull out, 'those ethnic wars would start again. I am absolutely sure of that,' he says. The result could be bloody. 'I would say the Ijaws have armed themselves to the teeth today. They would have the upper hand,' he adds, with the ghost of a nervous laugh. 'No doubt about that.'

Omiyi goes a step further, saying that Shell had to help the soldiers because of its position as junior partner of the government in oil production. Shell is 'just the operator', Omiyi says, launching into a rather awkward defence of the company's position. 'To a large extent you cannot be seen to be completely not supporting what government wants to do, if what government tries to do is humane, has integrity and is honest,' he says. 'If we ever know any time today that the government wants to use this for an offensive type of thing, we will not do it. That's the truth. And we keep this message all the time.'

Yet one of the enduring difficulties for Shell and other companies is that their names keep cropping up in disputes in which the military has very clearly gone on the offensive. I ask Omiyi about the notorious case of a town called Odioma. I'd visited the place early in 2005, to find a community razed to rubble by the army, supposedly in pursuit of a militant implicated in the murder of some councillors from a nearby village. The two communities were in dispute over land ownership, a conflict apparently catalysed by a visit, sweet with the promise of oil and money, from a Shell-chartered survey vessel.

What sticks in my mind from my visit to Odioma are the fatalistic words of Daniel Orumiegha-Bari, head of the village. He had sat incongruously in an intact chair that was surrounded by debris and wrecked buildings. He was carrying a Shell corporate diary, in which he neatly made notes. Surveying the wreckage, he had reflected stoically: 'It's not bad.' A colleague of mine had responded gently, 'It looks bad.' 'Well,' the chief had replied, 'it's life.' Then he had broken into two lines from what he had said was a Bing Crosby song:

There was something that was bound to happen and it happened somehow,
And now that something happened, it doesn't matter now.

Omiyi says Shell had 'nothing to do' with the Odioma attack. Shell had bought land from one of the communities many years ago, but another village had claimed to be the owner. Shell had left it to the government to oversee resolution of the dispute. 'What happened on the ground we don't know,' Omiyi says. 'I can't speak for the military.'

For all Omiyi's explanations, few people would deny that the oil industry's entanglements in Nigeria are of a dangerous complexity. This was confirmed most spectacularly in an extraordinary confidential report commissioned by Shell itself, which stated that the company was an 'integral part of the Niger Delta conflict system'. The document, prepared by WAC Global Services, a consulting group, compared the Delta with other 'high intensity conflict' zones such as Chechnya and Colombia, where more than 1,000 people are killed each year. Completed late in 2003, the report questioned Shell's ability to continue working onshore in Nigeria. It continued, 'The criminalization and political economy of conflicts in the region mean that the basis for escalated, protracted and entrenched violence is rapidly being established. This not only threatens [the oil industry's] future ability to operate, but also Nigerian national security.'

It was not long before Shell's rival Chevron produced a similar diagnosis of its own. Once again, the strength and directness of the language would seem extraordinary to anyone used to the bland statements about corporate responsibility made by multinationals in the annual reports issued to their shareholders. In May 2005, Chevron published advertisements in Nigerian newspapers describing its social policies in the Delta as inadequate, expensive and divisive. This very public self-flagellation confirmed what campaigners had been saying for years. Chevron said its system of showering some villages with money and projects had left other areas feeling 'alienated and underprivileged', leading to or fuelling conflict. Overall, it said, an unhealthy culture of rumour, blame and dependency thrived.

Chevron's conclusions left little doubt about the depth of the region's problems and the extent of the oil multinationals' involvement in them. The company said at least $500m of its property had been destroyed during the 2003 crisis. 'For the most part,' Chevron said, the Niger Delta was 'characterized by an unhealthy, unsustainable relationship, prone to conflict and division'. Violent, incessant fighting had left many of the development projects funded by the company destroyed, 'not to count the many lives and private property lost'.

Among the many failings identified in both the Shell and Chevron reports was the damage done by labelling the places where the companies operated 'host communities', entitled to benefits such as infrastructure, employment and building contracts. These advantages were denied to neighbours not so designated. The effect was to create divisions, leading to a potential for dispute between people who were not necessarily antipathetic previously. In other words, the companies created not only the opportunity for conflict, but new lines along which it could be drawn. 'You have these islands of people that are listened to, and then you have gaps in between,' Omiyi says. 'And most of the trouble we had came from those people.'

Shell and the other companies are now finding that there is no easy way back from a situation in which they have acquired a popular reputation for favouritism and bias. The company's official policy is to give a wider range of communities contracts for work in maintenance and other areas. The problem is that these deals tend to be hoovered up by local business magnates, who then leave to enjoy city life in Port Harcourt or elsewhere. So the wealth that is supposed to feed into the villages ends up melting away.

Another problem is that oil company funds for community development – which, in the case of Shell and its joint venture partners, totalled well over $100m in 2007 – are notoriously prone to corruption. Omiyi says the company takes the rooting out of graft seriously, adding that it sacks between 15 and 20 people a year – and about twice that number of contractors – for involvement in it. The number has not been going down, which disappoints him. It reflects a national problem, he says, as a culture has taken root in which people 'bribe for everything', including university places. 'Our staff are drawn from Nigerian society,' he says, stretching his hands out wide as if measuring the size of a fish he has just caught. 'So we cannot be a 100 per cent clean island in a very openly corrupt environment.'

Once again, Omiyi shifts from a small concession of Shell's short-comings to a much bigger criticism of the official failings that he says underpin the Delta's social problems. The region's crisis has indeed got worse, as predicted by the Shell-commissioned report, he admits, but this is because government is still not investing in the region. He tells a story of a conversation during which he urged a senior oil official to do more to improve security and living standards in the Delta. 'He said, "Come on Basil, you have been here for more than 50 years, you are going to be here for another hundred years. We will resolve this thing,"' Omiyi says, stopping for a knowing laugh. 'So people always push these things into the future. But that's the attitude.'

Omiyi knows Shell needs to show evidence of big improvements soon. The Delta's troubles are now so severe that they have begun to harm the industry's commercial position, despite oil prices spiking to record highs in 2008. Shell's financial model is to produce high volumes by taking advantage of low production costs, made possible because the 'light' and 'sweet' oil is both readily at hand and easily extractable, unlike the more viscous Arabian Gulf crudes. Shell needs to maximize production because, under its agreement with the government, the company keeps only a small portion of price increases above $50 a barrel. In other words, if you can't produce Nigerian crude in large quantities – which you can't when militants are taking over your fields and villagers hacking open your pipelines – you find it hard to make the top dollar you are looking for.

In an internal memo circulated in November 2007, Omiyi warned that Shell faced a 'grave' situation in Nigeria. He said the company's cost base was 'unsustainable' given the production outlook. He highlighted the chronic problem – facing all the multinationals – of the government's failure to honour its funding commitments to the production joint ventures. The tone was downbeat throughout, although Omiyi said he believed a restructuring programme due to be put in place by 2008 would turn Shell's fortunes around.

Shell and its fellow Western multinationals have faced further pressure from the increasingly blunt warnings issued by a government that felt emboldened – at least until the global financial crisis – by oil price rises unparalleled in history. Ojo Maduekwe, Nigeria's foreign minister, told the *Financial Times* in 2008 that Nigeria had 'woken up from sleep' and that standards not acceptable in Britain, the Netherlands and the USA would not be tolerated any more.

Maduekwe warned that the resource-stripping Western greed of the 1950s was no longer permissible in Nigeria; his country now demanded more in return. 'We want to remind those who are interested in oil and gas in 2008 that they were those who were interested in our cocoa and groundnuts, in 1958, and we don't see much difference,' he said. 'We think that our friends should now look at Nigeria, not in terms of how much more oil they can take out, or gas, but how much they can add to the value-chain.'

Omiyi insists these are not signs that Shell is on the way out in Nigeria. One reason is that the company has many fine oil and gas fields in an established production area where there is plenty of scope for further growth. He is bullish about the company's prospects of seeing off competition from Russia, India and China, none of which has yet put a 'dime of investment' into Nigeria. The Chinese state may be able to offer Nigeria much-needed infrastructure projects in areas such as railways, but Shell stands ready to expand its own contribution in sectors such as power generation. In other words, rather than loosening its clinch with government, Shell plans to tighten it. 'We are the strategic investors,' Omiyi says. 'It's not those guys.'

Yet, for all Omiyi's apparent confidence, Shell and its Western competitors in Nigeria increasingly appear caught in a vice. On the one hand, they are finding it increasingly hard to maintain production; on the other, they are heavily criticized internationally for their part in Nigeria's crisis. Asked if he thinks any of the criticism of Shell is fair, Omiyi says it is 'very difficult to answer that in general terms'. He offers a limited *mea culpa*, saying that perhaps the company should have moved into power generation earlier as a means of helping to make the Delta more liveable for its people.

On the company's wider role in Nigeria's troubles, however, Omiyi is bluntly unapologetic. There is no 'silver bullet' Shell could have fired to avert all the Delta's problems years ago. 'I haven't had any sense of tension by being Nigerian and being Shell,' he says of his own role. 'All I have actually gotten from that is a lot of respect from society.'

But surely, I say, Shell has been an integral part of an economy and a country that have simply not worked for their people? Again, Omiyi disagrees. People forget how backward the nation was at independence in 1960, he says. When he was a student in the mid-1960s, it took three days to get from his home east of Lagos to university in the western town of Ibadan. 'In whatever – health, social infrastructure, emancipation of people, I think Nigeria has done well,' he continues, although he adds, risking no overstatement, 'I think Nigeria could have done better.'

What has held the country back compared with others, he suggests, is the lack of structured development. The country as created by its British former occupiers was a 'huge compromise' of different peoples, cultures and religions, brought together as a state for only a brief period before they were granted independence. 'Our people have accepted mediocrity, some mediocre solutions, just to be able to keep everybody happy over the period,' he says. 'And that has not allowed Nigeria as a country to do the right thing.'

Critics of the oil industry argue that it has prospered these past 50 years in part precisely because of this constant national need for compromise and a squaring of interests. In such a febrile environment, there is a clear attraction for government in doing business with companies that offer you an easy way to tap your greatest resource, albeit at a price. It's not an organized corporate or Western diplomatic conspiracy, but rather flows naturally from events and decisions that in some cases date back to the colonial era and beyond. The power dynamics of the Delta have their roots in an era before most modern-day militants were even born.

For all Omiyi's professed optimism and facility for exculpation, there is a sense of foreboding underlying much of what he says. Towards the end of our meeting, he tells a revealing anecdote about what happened – or didn't happen – in 2008, to mark the 50th anniversary of Shell's first oil shipment from Oloibiri. The birthday of the resource that has driven so much of the country's life since independence was not marked with the kind of pomp and circumstance Nigerians are delighted to invoke for far lesser occasions. The government instead organized a low-key conference in Abuja, the capital, where Omiyi gave an 'introspective' historical talk. There were, as he puts it – in an echo of Oloibiri's Chief Inengite – 'no big celebrations' of an event that defined an epoch for Shell and for Nigeria.

Omiyi's downbeat assessment is a far cry from what the world had been saying about Nigeria during the preceding years, especially once the 2005 G8 summit promised its Western audience a development breakthrough on the African continent. British ministers had queued up to congratulate Abuja on a political and economic change programme that – for all the ability and apparent sincerity of some of the people involved – was seen by many critics as dangerously shallow and riddled with double standards. Hilary Benn, international development secretary, said the reforms on corruption and transparency were sending out a 'powerful signal' to other countries. When I asked a British official in Nigeria why London was making such uniformly positive statements, he told me patronisingly – and cynically – that the Nigerian media did not 'respond well to nuanced voices'.

As I watched the Live 8 music jamboree and the charity appeals around the summit, I began to wonder if they referred to the same continent I had lived on. Instead of pictures of vibrant, combative and messy Nigeria, here were images of dependent Africans whose salvation lay in the West's gift. The BBC ran a long drama by Richard Curtis, writer of *Four Weddings and a Funeral*, about an idealistic young woman who meets a senior British government official by chance in a café and ends up persuading the leaders of the G8 leading economies to do the right thing by the continent. It may have worked as a crude clarion call for people to do *something*, but it singularly failed even to nod at the many political and practical complications that surround attempts to help a country like Nigeria to right itself.

Most fundamentally, it ignored the fact that the problems in Africa's most populous nation were less about winning more aid than about making sure the money already in the country was being used properly. While the Western public was being told that a donation of a few pounds could save an African child's life, billions of dollars in Nigeria's greatest ever oil boom were still melting away in corruption and waste. Between 2004 and 2007 alone, the International Monetary Fund estimates the country's oil revenues topped £75bn, dwarfing the G8 nations' much-praised 2005 commitment to increase aid to Africa by $25bn over the following five years. Even as Nigeria was receiving debt relief from the Paris Club worth about $1bn a year in saved repayments, the oil industry estimated that even larger sums of crude were being stolen from the Niger Delta. Debt relief was morally right as a corrective to a historic injustice, but it was no more than the easy first part of a difficult and time-consuming solution to Nigeria's troubles.

As the Nigerian government celebrated the debt relief in late 2005, I spent a depressing day in Rivers State, one of Shell's biggest areas of operation. In Port Harcourt, I visited Community Primary School One in the Rumueme district, where head teacher M.C. Anwuri showed me her dilapidated empire. Many of the children didn't have desks and had to defecate on waste ground because of the lack of a working toilet. In the centre of the courtyard, three girls of about 13 were using bricks to bash charcoal against a patch of concrete; the end product was to be smeared on the wall of the classroom, to serve as a makeshift blackboard. In Anwuri's office, a calendar produced at no little expense by the state government told – or taunted – her that she had the good fortune to be living in the 'Treasure Base of the Nation'.

After I left the school, I drove across to the old Rivers State Government headquarters in Port Harcourt, for a meeting with one of the men who held the purse strings of the Treasure Base. My host, Tamunosisi Gogo Jaja, leader of the state's ruling People's Democratic Party MPs, was welcoming and brimming with charm. When I asked him for some documents, he banged on the thin plaster wall that divided his office from its anteroom. 'Love!' he hollered, flashing me a smile as if savouring the double entendre of his secretary's name.

A 38-year-old former publisher, Jaja described himself as the 'house leader' of the state assembly. The reason for his title, he explained, was that the People's Democratic Party held all 32 of the state parliament's seats. In other words, the party *was* the house, and vice versa. A poster on his wall showed the PDP state governor, Peter Odili, standing at the top of some steps, wearing robes of messianic white. Together, Odili, the state assembly and the region's ruling-party-dominated local governments controlled about £500m a year of oil money, about twice the entire national annual income of Liberia, a little further down the West African coast.

When I asked Jaja what the Rivers State administration had done with all its money, the first project he cited was the rebuilding of the road from the city centre to the airport. On social spending, Jaja said children under six and people over 60 were treated free at any government hospital. On education, he claimed there was 'no school in Rivers State that hasn't got well-furnished classes in terms of seats for the pupils, and tables and chairs for the teachers'. At this point, an activist friend who was with me intervened to tell him about our morning visit to Community Primary School One. She pulled out one of the day's newspapers, opening it at a long and highly critical newspaper article about schooling in the state. Jaja backtracked swiftly, saying that there were about 3,000 schools in the state, and that the government had had only six and a half years to improve them.

Now on the defensive, Jaja reeled off a revealing list of the government's other spending burdens. It had invested in a new set of buildings for its officials, including the governor and state assembly members. Jaja talked witheringly of the brick house built by the 'British colonial masters' in which the governor now lived. It would be a 'giant step' to have a Government House befitting of the administration of Rivers State. Another big investment was an air ambulance – which was sometimes used to shuttle state officials on trips around Nigeria. In 2006, the state authorities allocated Government House more than £20m to buy two helicopters and build a new airport runway and hangar.

The Rivers State Government had done a lot for its money, Jaja concluded. It would benefit from receiving even greater funding. 'We are even asking for more, truly, truly speaking,' he said, 'so as to be able to cater for the needs of our people.'

Jaja agreed to take me to see the almost-finished new State Assembly building. As we walked out of the existing headquarters, he gestured towards the chamber and commented on how small it was. He took us to the new parliament in his own car, a black Mercedes, which he weaved around the large potholes in the driveway. He had a smaller car – a Peugeot 504 – for official business. We talked about his family: his elder child, a six-year-old boy, had just started going to school in Rivers. He was, not surprisingly, being educated privately.

As we drove, Jaja described the Rivers State legislators' vision for their new headquarters. They had made 'serious contacts' with the British House of Commons and the US Congress, he said, to learn from them how to build an imposing home for lawmakers. In other words, the state wanted not just to ape the rich world but to surpass it, by putting its regional legislature on a par with the national parliaments in Washington and London. It was part of laying a 'solid foundation' for Rivers.

The new complex was already well on its way to completion. The blue-and-white superstructure was built, air-conditioners installed on the outside walls and a TV satellite dish fixed in place. The sourcing of building materials would have done the United Nations proud: there was white Italian marble, tiling imported from the United Arab Emirates, and Chinese mirrored blue glass for use in the central dome.

Jaja didn't seem worried about how I might perceive this big spending. As I talked to the Turkish building contractor over-seeing the building project, the house leader was already rolling his Mercedes towards the exit and another appointment. He gave me a friendly goodbye wave and drove off. He was so open and care-less about what was happening that I found it both ludicrous and horrendous. He did not even try to hide the official excess; for him, there was nothing either wrong or out of the ordinary about how he and his fellow politicians were spending the state's money. Maybe he calculated that a British national would be impressed by Nigeria's attempts to match or even better Western opulence and vainglory.

Even I was stunned by this open profligacy of officialdom during an era of supposed change. As central government was banking oil windfall billions in a special account, here were politicians in the Delta spraying their share of the cash around. Schools were crumbling even as official palaces were being built. Nations had revolted over less.

The full scale of the mismanagement was clear in the draft official Rivers State budget that Jaja had been unable or unwilling to provide when I'd asked him for it. Flicking through it, I was not surprised. It showed how misleading his proud citations of health and education projects were. Whatever initiatives were actually happening, the amount of waste and unaccounted-for spending in the government's own records revealed how much more could be done.

The budget had a contemptuous, 'let-them-eat-cake' tone, set in a preamble by Governor Odili. He said that official buildings such as Government House, the State Secretariat and the House of Assembly had become a 'source of embarrassment to all patriotic Rivers men and women'. All were being renovated; the Government House would be 'second to none in Nigeria'. 'We believe,' Odili said, 'that these projects will give our people a sense of pride in their institutions, restore our dignity as a people and make future governance more convenient, and effective.'

The governor's words showed little awareness of the contrast between the generous way his administration treated its own staff and the ruthlessness it applied to poor people. In Abuja, the Rivers State authorities were expanding their office complex and building 25 new official flats, including 10 three-bedroom apartments. In Port Harcourt, the administration was making people homeless by bulldozing shacks and businesses that had allegedly been built illegally. Odili described this as a 'worthwhile initiative'. 'Enforcement of discipline has continued as [a] deliberate effort to keep the city of Port Harcourt clean,' he said.

But discipline and belt-tightening had their limits. Spending on Government House had risen by almost a third and now accounted for almost 5 per cent of the state budget. The Government House gardening bill had more than tripled in a year to £20,000, while the swimming pool costs had increased almost 40 per cent. Newspaper and magazine costs had multiplied four-fold, while 'staff uniform and welfare' costs had increased from a modest £6,400 to a hefty £200,000, or about £600 for each of the complex's 334 employees. Most strikingly of all, the cost of 'gifts and souvenirs for visitors' had risen from £200,000 to £1,000,000: in other words, a state whose people mostly lived in grinding poverty said – if we take it at its word – that it was blowing a million bucks a year on trinkets for its guests.

Peppered through the departments, large and small, were allocations for vehicles, mainly Mercedes four-wheel-drives and Peugeots. Government House's 'new vehicle' bill alone, which was not itemized, came to £3.2m, equivalent to 228 Peugeots or 69 Mercedes 4x4s. The Ministry of Justice had a larger allocation for cars than for law books and journals. Elsewhere were scattered allowances for air-conditioners, fridges and, in the case of the culture ministry, almost £10,000 for installing and subscribing to satellite television.

The deeper one went into the book, the more mysterious and greater the costs, as if the compilers were hoping readers wouldn't get this far. Tucked away on page 300 were Government House security costs: these were estimated at £14m, up 75 per cent on the previous year. 'Unspecified special projects' – described in exactly those words – would cost £18m, up from £7m the previous year. Unexplained 'investments' would cost a further £400,000. It is almost as if the authors were making up new categories because they had more money than they knew how to spend.

Near the end was perhaps the most poignant entry: it was the spending on horticulture in Port Harcourt, which had once been known as the 'Garden City'. An annual budget of just £60 – less than 1 per cent of the Government House gardening bill – was set aside for grass-cutting in Isaac Boro park, the main public green space. The same amount was given over as the entire allocation for flowers in the city. As a statement of a ruling class's sense of history, civic pride and attitude to its electorate, it seemed as eloquent as any of the government buildings and shiny official cars brought into being by the soaring oil price.

The Rivers State budget was one of many documents and events that chipped away at the credibility of the reform story being promoted by Britain and other Western countries. At Obasanjo's inauguration in May 2003 for his second term as president, he had spoken compellingly, even movingly, about how he could not accept that Nigerians would, 'in preference for a decent and civilized society, opt for one in which law and order are disregarded and regulations are circumvented as the norm'. Yet even the president's own propaganda suggested there was a considerable gap between his dreams and what was happening on the ground. In a 2003 edition of *Villascope*, a magazine published by the presidency, a feature on the police inspector-general, Tafa Balogun, described him 'leading the war against crime'. In 2005, the same Balogun was sent to jail after being convicted of charges relating to the alleged theft of more than $100m of public money.

This ambivalence shouldn't have come as a surprise to anyone who had noted the ambiguity that has defined Obasanjo's career. By any standards, the former president has had one of the more extraordinary public lives of the last century, taking on the roles of civil war general, three-time head of state, agriculture magnate, prisoner of conscience and international peacekeeper. Yet he has been an integral part of the Nigerian élite power nexus for more than 40 years, during which time corruption and human rights abuses have flourished. It is a nuanced legacy that led perplexed Nigerians to ask me again and again: Why did Obasanjo and his government get so much praise abroad when they were frequently looked at so much more sceptically back home?

London seemed more interested in perpetuating the perennial – and seemingly pathological – Western habit of painting African countries with a crude roller instead of a fine brush. It is as if the rich nation politicians, journalists and NGO officials who analyse supposedly reformist African nations are afraid to allow their rulers the dignity of being written about as real people, possessing flaws as well as flair. It is a peculiar kind of reverse racism, with Westerners treating 'progressive' Africans like china that might shatter if subjected to the normal wear and tear of political debate.

Those who raised questions about the Obasanjo administration's agenda risked being tagged cynics by both London and Abuja, incapable of either seeing change or overcoming their prejudices. But those who claim to be optimists can be cynical, too, particularly when – like the two governments – they share an interest in projecting an impression of rapid improvement in the state that has styled itself the 'Heart of Africa' and is key to Western energy security. I always found it revealing that – for all the achievements and bravery of a handful of Nigerian officials – few people outside government seemed to think reform was going far enough or fast enough to pull society from its oily mire. Most strikingly of all, by the time the president left office in 2007, events in the Niger Delta oil heartlands were making the international case for a Nigerian renaissance look at best foolish and at worst disingenuous.

9

NOT HOSTAGES BUT JOURNALISTS

I am sitting sweating in a car pulled up outside one of Port Harcourt's waterside slums, nervous and impatient for the day to begin. I barely register the crowds passing close to me on this narrow road as they head to work or to perform household tasks. My taut nerves are stretched a little further by the sound of a loud bang on my vehicle's bonnet. When I look up in alarm my driver, Kunle, points in wordless explanation at the mango tree whose fruit-heavy branches are hanging over us.

I've come early this close tropical morning to wait for three men who have promised to take me to the heart of the Niger Delta's main – and deadliest – militant movement. I feel the same unfocused tension I'd experienced on the Abonnema jetty, waiting for Asari's men three years previously. This time I am due to meet members of the Movement for the Emancipation of the Niger Delta, or MEND, the militant group that sprang up while Asari was in prison. The organization had attacked oil installations and kidnapped dozens of oil workers, prompting big companies to send non-essential staff home and shut down hundreds of thousands of barrels a day of production. For the oil majors, long used to a bit of heat, the security crisis was as bad as they had known.

From a distance, MEND seemed to me an enhanced and more sophisticated version of Asari's group, a sort of militant version of a computer upgrade. It had learned from his mistakes of being too open, too easily targeted and too careless in his statements. Its members wore masks, kept the location of their bases vague and avoided putting forward a single leader who could be picked off. They gave the impression of being well organized and serious, posing masked for pictures in which they pointed their guns at a hostage. They showed a deep understanding of how the international media and politics work, calibrating reassurances that the Delta was not Iraq or Afghanistan with warnings of further attacks on the still mighty but increasingly beleaguered oil industry.

The rise of MEND showed how the troubles of a country trapped in a terrible past were increasingly becoming the world's intractable problems, too. MEND's name was becoming more and more well known to international oil traders, but the organisation remained enigmatic, at times seeming more of a mood than a formal movement. Its members would pop up to plant a bomb or take a hostage, before disappearing back to the creeks. The guerrilla tactics were highly effective against foreign oil interests that were now being confounded by the very Deltan turmoil their own behaviour over decades had helped create.

At first, it had looked as if my attempts to meet MEND would fail. The group's self-proclaimed leader, who calls himself Jomo Gbomo, had kept a playful distance from my attempts to arrange a visit. Gbomo is another strange emblem of the Delta's Byzantine conflict. Media-savvy and eloquent in his prose, he knows the value of elusiveness. He declined to see me, saying that he met no journalists and operated only through an e-mail address. He was sure, he said – in a phrase more suggestive of a British public school than the swamps of Nigeria – that I was already quite familiar with the Delta's 'sob story'.

After a week of fruitless cat-and-mouse games with Gbomo, a freelance journalist I knew with vast Delta experience, Glenn McKenzie, had come up with an alternative plan. He knew some people who knew MEND militants and could help us meet them, without Gbomo's assistance. I had to make the initial contacts and the arrangements for a trip.

The previous day, I'd called the key man, who was known simply as Comrade. He had asked me to come to see him at one of downtown Port Harcourt's many slums. There, I was met on the road by one of his foot soldiers, a young man called Martins, who led me into the settlement, through a network of filthy tracks flanked by dense, claustrophobic, one-storey shacks. As periwinkle shells crackled under my feet, I was reminded of my trip to Oloibiri, the restless place from where Nigeria's first oil was shipped to Britain. As if reading my thoughts, Martins asked me a pained rhetorical question, 'You see our roads in the Niger Delta?'

After a few minutes, we arrived at a large building that was little more than an open-sided wooden frame, like the skeleton of a barn. It was perched on a spit of land surrounded by water: truly a gateway to the creeks. A welcome breeze disturbed the humid stillness, but it was still hard to escape the pervasive slum stench. A group of young men were sitting around, some of them playing draughts. One of them told me the smell was from a dog that had died recently on one of the tiny islets dotted yards from the shore.

In one corner of the camp was a shrine, dominated by a flag that depicted a map of the Delta, a big black star and an oil well. It was the emblem of an organization called the Movement for the Survival of the Ethnic Nationalities in the Niger Delta, or MOSEND. Over the years MOSEND had evolved from a group focused on the interests of the Ijaw, the Delta's largest ethnic group, to a pan-ethnic movement aimed at promoting resistance across the region. It was one of a network of groups MEND had tried to tap into.

Once Comrade arrived, I immediately sensed the unusual presence that others had talked about. Physically, he was far from imposing, even a little odd. He was tubby, and had a large bandage taped to the back of his head. His voice was high-pitched to the point that my driver, taking directions from him on the phone, had mistaken him for a woman.

Yet, a little like Biafra's General Ojukwu, Comrade had the assured air of a man who expects to be obeyed and consequently is. The camp had an air of calm that I did not readily associate with the many gatherings of angry young men that I had seen in the Delta over the years. Comrade offered me a Maltina – a dark, sweet, stomach-filling malt drink – and we started to chat.

His opening gambit took me by surprise. He started by attacking his own fellow-countrymen for what they had done to the Niger Delta. People abroad respected Nigerians for their oil wealth, he said, but that was not the feeling among people locally. While foreigners had come to help the region develop, Nigerian politicians and Nigerian oil company managers were playing dirty on the ground. If the militants were going to take hostages, 'let them take black hostages', Comrade said. 'Let them take [Nigerian] directors, chairmen, senators, governors. They are the people who are causing the problem.'

Warming to his theme, he gestured to an older man sitting on the bench next to me. The man, Ernest, had been working as a chauffeur in the oil industry for 15 years, yet received no benefits such as hospital treatment. This showed the lack of compassion among the wealthy Nigerians who employed Ernest and other people in similar positions. One of Comrade's colleagues chipped in, 'They become greedy, the blacks.' Even my driver, Kunle, chimed in agreement, 'That's the problem.'

This conversation was yet another twist in the labyrinthine process of trying to understand what is happening in the Niger Delta. Here were ostensibly some of the toughest, fiercest, proudest opponents of the oil multinationals, yet they seemed to see their fellow Nigerians as the greatest obstacles to progress. At root, it seemed, Comrade didn't want to expel the multinationals, but to reach an accommodation with them that would allow this long-shackled country to develop at last. Comrade said he wanted foreign companies 'to trade with us, be friendly with us, to get us to other areas that other people have gone to.' 'We are trying to cut away,' he said, 'from armed struggle to industrial revolution.'

Then, just as quickly as he had offered this glimpse of reconciliation, Comrade had pulled back. He'd said his men were being propelled by circumstances into a battle they would rather not fight, but were fully prepared for. While he talked, I had spotted out of the corner of my eye a punchbag set up just outside the camp, perpetually ready to be beaten for training or in simple frustration. If the inhabitants of the Niger Delta were pushed to the wall, Comrade had said, there would be no compromise. 'The decision will be final,' he'd said: 'No retreat.' He was echoed by one of the draughts-players, who had looked up from the board and said simply, 'No surrender.'

As I continue my morning wait under the mango tree, I have plenty of time to contemplate both Comrade's words and the condition of the Delta. On the way from my hotel, my driver and I had passed a queue of vehicles spilling out on to the road from the forecourt of a petrol station run by the state Nigerian National Petroleum Corporation. The station was offering petrol at 64 naira per litre, compared with the 65 naira charged by the almost-deserted Conoil filling station down the road. In other words, drivers were prepared to queue in order to save perhaps 50 cents on a $25 petrol purchase. It showed how narrow were the margins fought over in the Delta.

When our guides finally arrive, we set off for Yenagoa in neighbouring Bayelsa State, the starting point for a journey deep into the creeks. At Yenagoa jetty, after linking up with Glenn, we hire a speedboat and cast off. As always on the trips to the heart of the Delta, I feel as if I am being propelled into a more welcoming world. A bracing wind replaced the humid closeness of Yenagoa – a town clustered around a single, thunderous main road. The foliage on either side of the water is thick and lush, with oil palms peeping over the top of the treeline like small children looking over a wall. The river traffic – mainly canoes loaded with goods such as fish, wood and plantain – clings to the banks to avoid being capsized by our wake.

Soon, we pass a village where a long white flag flutters from a post, just like the one I'd seen at Asari's camp. It's a symbol of Egbesu, the water spirit so central to Ijaw culture. Simeon, one of our guides, tells us that the white flags represent peace, the red, fighting spirit. Ijaw militants say they rely on Egbesu to protect them; yesterday, Comrade had asked if I could drink battery acid, or be shot at by a bullet – two things that Egbesu devotees are supposed to be able to do without harming themselves. If you are killed in battle, it means not that Egbesu has failed you, but that you have violated its laws. As Simeon puts it bluntly, 'You oppress, you steal, you will die.'

Simeon initially seems an affable, avuncular and unlikely fixer to help us meet a notorious armed militant group. He is thin, wiry and wears bifocals tied together with a piece of string: his third pair of glasses, he says, after he lost one and dropped the second in a creek. He has a salt-and-pepper beard, and teeth as jagged as a mountain range. His appearance would be professorial, were it not for the troubadourish bagginess of his billowing yellow traditional smock and trousers. He looks to me like a man who has made a living out of being underestimated.

After about an hour on the river, the boat driver suddenly cuts the engine. Our guides all raise their hands in the air, prompting Glenn and me to follow suit. We are approaching a checkpoint of the Mobile Police: they are notoriously trigger-happy at the best of times and are likely to be especially skittish at this volatile time.

Two officers beckon us to a jetty, from where we clamber up to meet them. They ask Glenn and me a series of questions: Is the state government aware of our visit? Where is our security? What did we think about the 2007 national elections held the previous week? We evade and lie, saying we have all the permissions and are very impressed with the democratic process in the Niger Delta. They seem satisfied with these thin answers. The whole process has a slightly half-hearted feel, as if their main purpose is to cover their backs in case anything happens to us.

After a ride of another hour or so, the cloudy and cool weather begins to change for the worse. The wind strengthens and it starts to spot with rain. Suddenly, a speedboat appears from a tributary to our left and speeds across our course. Looking to our right, we see other people: a line of men on the bank, all completely naked. About 15 of them are waiting to go into the water, where half a dozen others are already splashing around. It is our first, arresting, glimpse of MEND fighters.

The other speedboat pulls up alongside us and a young man gets out to greet our guides. One of them, Godson, produces a bottle of Chelsea gin, from which he has been taking nips during our journey. Our visitor takes a swig, then shakes a spray of liquor four times to the left of our boat's stern, as if in blessing.

The new man introduces himself to us as MEND Commander Timi Freeman, although his dress seems more suited to a day on the golf course than a militia war. He wears a visor and a mauve-and-white Breton-style horizontally striped top. But instead of a string of onions hanging round his neck, he has a Nokia headset.

Commander Freeman is a youthful-looking 30, his face still filled out with some of the puppy fat of youth. His nickname among the other militants is, rather sweetly, 'Fineboy'. He tells us we are now in a zone under the total control of MEND, where the government soldiers garrisoned in the area dare not approach. 'They can't come here,' he says. 'They fear. We don't look for trouble. But if the military happens to come here, they are looking for trouble.'

Under Commander Freeman's instructions, we change course and take a sharp left down another branch of the river. Some minutes later, we moor at the jetty at the village of Korokorose, where the children welcome Glenn and me by breaking into a chorus of '*Oyinbo Pepe!*' the traditional kids' greeting to white foreigners. A young man carrying a rifle and wearing a black T-shirt depicting a Hummer – the military-chic four-wheel-drive vehicle beloved of wannabe warlords – meets us. He leads us to a half-built house opposite a field of tall grass, where we are asked to wait for the field marshal.

The field marshal arrives and sits across from us, his face half-obscured by his hat. He listens to Simeon's introduction and then offers us a drink. I take a non-alcoholic malt; many of the young men take Star beer. We sit and wait for the field marshal to speak, but he is as taciturn as he is hospitable. Soon he gets up to go; we are told we will see him again later, but we never do.

It's clear we will have little choice in determining the course of the day's events. We go back to the boat and are whisked away to a village called Ikebiri I. There, we wait again, this time in the darkened room of a dilapidated house, the dried mud on its walls crumbling away. Outside, a huge crowd gathers despite the rain, with children seemingly spilling in through every door and window. One boy, framed strikingly in the door by the light, has the horribly distended stomach that indicates kwashiorkor.

Eventually, the boy stands back to allow an older man to enter the room. After introducing himself as Francis Ododo, he launches almost straight away into an account of how militancy in the Delta is fuelled by the people's feeling that they have been robbed. He attacks Olusegun Obasanjo, the military leader turned civilian president, over his 1978 Land Use Act, which gave the federal government ownership of all the region's oil. 'Now the Niger Delta man has seen that he is nowhere,' Ododo rages. 'And that's why any problem that you are seeing today is happening.'

Ododo uses Nigeria's new banknotes as an illustration of his thesis that the Delta has been marginalized. These smart bills are long overdue in a country where small notes often seem close to dissolving in the accumulated moisture of the scores of sweaty pockets through which they have passed. The new notes feature pictures of notable individuals from various parts of the country. But this attempt at national balance is double-edged: in a nation of hundreds of ethnic groups speaking hundreds of languages, you are bound to leave some-one out. So it is that the 50 naira note features writing in English, Igbo, Hausa, Yoruba and even Arabic – but not Ijaw. 'Where is the Niger Delta man?' Ododo asks rhetorically.

As Ododo continues, I am struck again by how the Delta's prob-lems are linked not simply to poverty, but to a rising consciousness of it. As travel and television become more widely available, people from the region are increasingly looking outside and seeing what they have been excluded from. 'You have the youths who have the future in their hands, and they are also travelling out to other parts of the country,' Ododo says. 'After seeing how those areas are developed . . . the youths can't bear it again. People from far off are benefiting from our resources.'

The result is that many longstanding oil-producing communi-ties such as Ikebiri I are becoming increasingly restive. People are unhappy, Ododo says, because Chevron, which operates in the area, still flares its waste gas, causing acid rain and polluting the area with night-time light. Chevron doesn't employ enough local people, and neither it nor the government have brought development: petrol in Ikebiri I, a prime source of oil, costs 150 naira a litre, or more than twice the going rate elsewhere. There are no roads linking Ikebiri I to the main cities, so a trip to Yenagoa costs a prohibitive £10 return. In an arresting echo of Comrade's rhetoric the previous day,

Ododo talks about how Niger Delta youths have been 'pushed to the wall'.

Not for the first time in the Delta, the critique is making me uncomfortable. As a white incomer, I often have the sense of playing dual, even contradictory, roles. I am a potential messenger for people's frustration, and at the same time a symbol of the very privileges they say they resent. It is a conflict that I feel perhaps more strongly than many of those I talk to. For them, the priority is to tell their story to a world that they optimistically hope is listening.

At a deeper level, I am also aware of how the Delta has, over the past few years, become what a friend describes as very 'Hollywood'. Many villagers and militants, sensing the growing if still modest international interest in the region, operate with a sense of theatre. Local people have an increasingly sophisticated understanding of how the international media works and how it can be used to serve their interests. I am reminded of this as we pull away from the Ikebiri jetty, when Commander Freeman gestures to a house on stilts at the river's edge. 'See where we toilet, see,' he says, urgently. 'See the houses that we are living in. Can you believe we are in an oil-producing community?' A moment later, having paused for thought, he turns to us to say that next time we come we should 'bring video coverage'.

En route to our next stop, Freeman and I chat about his life, which has been shaped since his teenage years by conflict. He tells me he has been in four militant organizations since he left school in 1994, including Asari's Niger Delta People's Volunteer Force. As he talks, I notice two scars on his hand, one long, the other in the shape of a cross. He says they are injuries caused by a soldier's machete.

In 1999, Freeman says, he was sent to prison after a shootout in Port Harcourt in which the military killed two of his men and he killed a soldier. He suffered greatly during the five years he was held without trial, but he still decided to rejoin the militants once he came out. None of the underlying injustices he was fighting about have changed. 'I became a militant because of the spoiling of my people,' he says. 'We have oil but no development. That is why I have said I will fight until my last bullet – until whoever will kill me.'

When I ask him if he has a family, he smiles and replies that he has a mother and father. Is he married? I ask. No, he replies. Would he like to get married? He smiles charmingly, gives me a thumbs-up sign and says, 'Yes.' I say that it must be difficult to find a wife when he leads the kind of life he does, alternating between a week in the camp and a week in Port Harcourt. He shrugs and says he sees no choice. We fall silent, and after a short time he taps me on the arm gently. 'Before you marry,' he says, 'you have money. I have no money. If I have money, I will marry.'

It is a gentle insight into the straightforward desires of a man who claims to have killed more than 30 soldiers in an adult life spent alternating between war and jail. Freeman says his territory consists of three camps, each of which house between 100 and 200 men at any one time. The militants are armed with rifles, light and heavy machine guns, and even two rocket-propelled grenade launchers. One of these they bought, while the other they seized from government soldiers. Asked who supplies the weapons, he laughs, a little coyly. 'We buy them from the military.'

Sometimes arms come from neighbouring Cameroon, he says, sometimes from further afield such as Liberia, a country wasted by more than a decade of civil war. He claims that the weapons dealer they regularly use even has a supply line from Iraq. Asked where the men get the money to buy the arms, he smiles and thinks for a time, as though weighing up how much to tell me. He finally replies that the militants take oil from the pipelines to sell on the black market. In other words, just like Asari's men, they finance themselves by exploiting the very oil industry that has blighted their region for half a century.

By now, we have travelled to the very limits of the Niger Delta, where the river meets the sea. In front of us, through a wide gap in the sandbanks, I can see the waves of the open ocean. Further back still, I can see the gas flare of an offshore oil field, silent witness to the chaos unfolding on land.

We veer away from the sea's jaws and put in at the village of Kuluama I, another area where Chevron operates. As ever, the local leaders have the statistics of the oil industry and its failings close to their fingertips. Emmanuel Orumo, chairman of the Community Development Council, tells us that there are hundreds of jobless youths, who are turned away from work by the company and then kept at bay by 'uniformed men'. 'We have graduates who go for interview,' Orumo says, adding that Chevron fails the community 'woefully'. The multinational has declined to comment on questions I put to it about the complaints in Kuluama I and Ikebiri I.

Kuluama I is another community whose relationship with its First World guest appears to have completely broken down. It has something else in common with many other Delta villages: terrible inconsistency in its choices of friends and enemies. One of the men sitting around while Orumo talks has a T-shirt that proclaims his 'total support' for DSP Alamieyeseigha, the disgraced former Bayelsa governor.

Orumo takes us on a further whistlestop village tour in our already breathless day. He deplores the lack of road and the rusting zinc roofs, which people link to acid rain caused by oil company gas flaring. We visit a blackened pool from which the local people take their drinking water. Someone plunges a pink plastic cup into the murk, fills it and thrusts it towards me, asking angrily, 'You see the water? Can you drink it?' As if reciting a mantra, Orumo tells us not to forget that oil was discovered in Kuluama I in 1963. In other words, while Britain was revelling in free love – and enjoying the first shipments of Nigerian oil – this Delta village was starting on a decades-long journey dominated by recrimination rather than liberation.

We climb on the boat again to be taken to Kuluama I's most alarming sight of all. A few minutes' ride away is a sandbar, where blackened mangrove roots twist as if in silent agony. A thin layer of seawater rushes in and out over the sand in rhythm with the waves. The encroachment, villagers tell us, is becoming greater and greater and will eventually wash the community away. They blame the problem on seismic activity by oil companies such as Chevron, whose offshore platforms are visible in the distance. It could equally be global warming, itself indirectly attributable to the oil industry.

People's fear and anger at this apocalyptic prospect is visible and genuine. Reuben Wilson Clifford, a local man, holds his arm in the air like a weapon, and promises destruction if the village is indeed washed away. Someone else shouts out, 'We will carry the gun to claim our rights. Since we have nothing – nobody cares for us.' It is an eerie echo of a conversation I'd had the previous week with an oil executive: he said the level of violence that had developed in the Delta was not surprising, as people there had nothing to lose.

Whatever the cause of the sandbar flooding, there is something unsettling about this place where young men rage under the oil industry's distant and baleful watching eye. As we stand surrounded by the withered plant roots, the sand sucks in my boots and an incoming wave submerges them. Truly, this is a place on the edge: as we leave, we pass houses on stilts surrounded totally by water, ready to be taken by the flood, if and when it comes.

By now, with dusk looming, it is too late for us to go back to Yenagoa. The creeks after dark are not the place to run into the army or mobile policemen. Glenn and I decide to take up the militants' offer to stay for the night, although we don't know where we are going or where we will sleep. We have long since lost control of this trip, if we ever had it in the first place.

After a breakneck ride back along the main waterway, we plunge into the mangroves. The creeks here are narrow and claustrophobic: the sea can't be too far away, but I'm unable to see enough through the vegetation to make out where. I can hear cicadas at first and then, more and more clearly, drumming and chanting. Godson tells me not to be afraid, as we are in God's hands.

Ahead, two men with guns are perching in the mangroves, their faces part-covered with cloths. The boat stops at a little gap in the vegetation next to them, through which we can hear the chanting nearby. Godson starts taking his trousers off, and Glenn and I are told to do the same. Then Godson and our fellow passengers step out of the boat one by one and into the deep creek water. They do not sink, and I see they are stepping on a dense but brittle cat's cradle of mangrove roots that lies beneath the surface.

By the time my turn comes, I feel close to passing out through a combination of nerves and hunger. I start to creep, slip and grasp my way towards the shore, willing myself to control my fear. Eventually, the riverbed becomes shallow enough to walk on. We wade in line to the shore, where a man is waiting to bless us. As each of us passes, he flicks water at us, then sprinkles a white solution known as native chalk from a bowl he carries at his side. 'You are holy!' he exclaims as he finishes.

A few yards back from the shore, there are about 20 men dancing to a beat tapped out by drummers, one of whom strikes an old car exhaust with a stick. As we walk towards them, I have a strange sensation of being simultaneously the centre of attention and completely ignored. All around, the air smells pungent with hashish and incense.

I look round at the other dancing militants, whose appearance blends the whimsical, the spiritual and the surreal. Many are in their boxer shorts; some of these bear Tommy Hilfiger labels, while one pair is decorated incongruously with pictures of babies and flowers. Some of the men wear white or red cloths round their faces, and one has a white sheet on his head with holes for the eyes, like a ghost. It reminds me of the cover photo of Franz Fanon's *Black Faces, White Masks*, the classic treatise on black consciousness.

Our companions from the boats immediately join in the dance, as caught up in the rhythm as they had been in discussions of Delta politics earlier in the day. Reuben Clifford, who had spoken with such force about Kuluama's fate, is even more pumped up now. Standing a foot or so away from me, he shouts that his men are going to continue to capture white people like me. He could tie me to a tree now, shoot at me and no bullet would enter, he says, because I have just been blessed with the spirit of Egbesu. He grabs my arm urgently. 'We are jobless,' he says. 'The best way is to catch you people, flog you. That is the way we live.'

In the confusion, someone shepherds Glenn and me towards a small man with a white cloth around his face. He says he is called Commander Three Lions. He gives a rapid-fire talk about the history of the Delta, going back to the days of the Royal Niger Company and beyond, but the speed of his speech and the noise around us makes it hard to make out much of what he is saying. My notes on the conversation reveal a stream of consciousness, perhaps mine as much as his:

*The whole Nigeria . . . the criminal governance . . . there is nobody
can take our oil . . . I fight Nigerian government . . . I am prepared for
death . . . you people go back and tell them . . . we are not fighting for
resources now . . . we are fighting for our lives.*

As Three Lions speaks, two of the dancers hold up a large piece of
white cloth between them. A third man steps a few feet back from
it, takes aim at it with his rifle and fires, making me jump. There is
a cheer, and the two men holding the cloth raise it to show that it is
unmarked by a bullet hole, as it has Egbesu's blessing. Later, I wonder
how he did it. He could hardly have shot a live round into the crowd
of dancers; yet if he had knowingly loaded a blank he would have
been acknowledging that the Egbesu magic didn't work.

Encouraged by the militants, Glenn and I briefly and nervously
join in the dancing. Then, as suddenly as we were propelled into this
intoxicating world, we are told we should leave. We scramble back
across the mangroves. 'It seems easier on the way back,' I say, seconds
before slipping and almost plunging into the creek. No doubt Egbesu
is punishing me for my complacency.

As we glide away from this secret militant garden, Freeman reflects
on an encounter that was perhaps even closer to calamity than it
seemed at the time. 'They wanted to tie you with one cloth and shoot
you,' he says. 'But I told them not to.'

On the way back to Korokorose, where we are to stay the night,
Freeman tells another story that reminds me that any spirituality in
the Delta's armed struggle is usually trumped by pragmatic material-
ism. He says the militants are unhappy with the Ikebiri I community,
which we visited earlier, because it is in dispute with Agip over an oil
pipeline that local people broke open deliberately. The community is
refusing to take 2m naira being offered by the company to allow it to
continue to work, but the militants want the villagers to accept. The
reason: the militants need the pipeline to be in use so they can take
oil illegitimately to fund themselves. Agip has declined to comment
on the incident.

When I ask Freeman if he thinks the militants are being fair to the
people of Ikebiri I, he shrugs and says that the villagers would benefit

if oil started flowing again. 'If we loaded up to five barges, we would give one to the community,' he says.

By now, the river is dark and we are navigating by a combination of the boat driver's torch and Godson's hand signals. We pass villages where the only light is from oil lamps and the occasional generator-powered electric bulb. Beyond them is a ghostly orange glow behind the tree line. It is the oil-company gas flares, burning constantly as if in some cruel experiment in sleep-deprivation.

By the time we reach Korokorose the boat is quiet, through a combination of our physical tiredness and immersion in our own thoughts. As we jump onto land once again, Willy, one of our guides, offers me an unnerving parting shot that echoes Comrade's observation about who is most responsible for the Delta's troubles. It is as if deep beneath the anger and ruthlessness of these militants lurks a sense of inferiority and even self-loathing founded on centuries of exploitation and grotesque power relationships. 'If white people like you had been in charge, there would be no problem,' Willy says. 'Because you are not dubious. But we Nigerians are dubious.'

Once ashore, we are asked to wait in a bar on the main street of the village, while our guides sort out somewhere for us to stay. Inside, young men are drinking beer and soft drinks to a soundtrack of loud music and a generator's hum. There isn't much chatter, partly because of the booming sound system, and partly because the arrangement of chairs around the sides of the room gives it a gladiatorial feel.

Glenn and I start talking to some of the young men, one of whom suggests kidnapping one of us and splitting the ransom 50–50. The question of ransom payments for oil workers is one of the great open secrets of the Delta: no one believes the official line that they are never paid. The previous week, I had asked an oil company executive how ransom demands were settled. He replied, 'No comment. But we do not pay ransoms.' When I asked if the government paid the ransoms for the multinationals, he said, with insouciance, 'Pass'.

My neighbour in the bar, crew leader Henry Emmanuel, says he has no job and has taken up arms to make things better. He says he doesn't know the international oil price exactly, but he knows lots of people are making a great deal of money out of Nigeria. 'When government does things the way we want, we will drop our weapons and follow our path into society,' he says.

The longer we stay in the bar, the edgier the mood becomes. A powerfully built man in a red T-shirt comes in, looks at us a little suspiciously and leaves again. Shortly afterwards, we hear the sound of an argument outside. Glenn and I go out, where we see the man in the red T-shirt berating Simeon. When we appear, the man turns his fury on us, despite Glenn's best attempts to placate him. It seems we and our guides have misjudged the terrain: we are not as welcome here as we had thought.

As we retreat inside, red T-shirt man follows, continuing to berate us. He shouts about how people are suffering, his wife is suffering, and he can't see his children because the police are looking for him. That's why he kidnaps white people, he says. He's been told we are journalists not spies. 'That is the only thing,' he shouts, 'that is saving you people.'

Glenn and I are no longer sure if it's a good idea for us to stay in the village tonight. We say to our guides that we are prepared to risk the run back to Yenagoa, but they assure us everything will be OK. Godson speaks to the angry man to try to calm him down.

Eventually, Glenn and I are shown out of the bar and to the house of a wealthy local man, where two armed guards are posted to watch over us. After some food – and after I have turned down Freeman's offer of a woman for the night – we hunker down with our guides at the house. Simeon is philosophical about the dispute we became involved in earlier. 'The grammar to explain this,' he says, with the glimmer of a twinkle, 'is adventure.'

The next morning, we wake early and wait a little nervously with Simeon in the front room. Outside, local people are cutting the grass with machetes. At about nine o'clock, a delegation arrives to ask us to come to see some of the community's leaders. We troop out with our guards, wondering what the many watching villagers are thinking. We are not prisoners yet, but we have an ever-sharpening sense of our own vulnerability.

We arrive at a house, where our antagonist from the previous night introduces us to the village rulers. His words are reassuring, up to a point. 'They are not the kind of white people we kidnap to talk to the government,' he says. 'They are not hostages – they are journalists. They are here to see how we are suffering and why.' He taps his gun. 'That is their mission. Everybody has to tell them their problem.'

The messages from the people of Korokorose are similar to those we have heard elsewhere. Increasingly, the members of the community contrast the development of Yenagoa and other urban centres with their own situation. Again, the community has access to just enough material wealth and services to show it that some great unfairness is at work. A.A. Paul, one of the local leaders, had begun the meeting by switching on the television news. 'Sometimes the Nigerian government takes pictures of Lagos to show people outside that Nigeria has developed,' he says. 'But when I come here I see the opposite thing altogether.'

The local leaders are already annoyed that an agreement they signed with the Italian oil multinational Agip last month has yet to yield any development. They show us the document, in which the company promises to supply a range of items over the coming years, from speedboats to a new science laboratory for school pupils. Like other such deals between communities and companies across the Delta, it is full of caveats that allow the company to rescind its responsibilities in the event of any unrest. It also offers the community scant prospect of redress if they feel they are being unfairly treated: the committee overseeing the programme has a majority of oil company representatives.

When we come out of the meeting, we find that we have caused yet more trouble. Freeman is annoyed with us for leaving the house without consulting him first, although we didn't feel we had any choice.

The argument raises the tension further as we start a tour of the village, during which we are keeping half an eye on what we are supposed to be looking at and half an eye on our own security. We see a local school that has become a concrete shell. Wood is breaking away from the tops of desks and chairs, revealing metal frames. Its condition seems to say to the children, 'You don't matter'. All the while, the militant who was angry with us last night is taking swigs of gin in the morning sun.

As we are hustled round the village, I wonder how many of the onlookers think we are hostages. We are certainly not kidnapped, but neither are we totally at liberty. I am busily instructed by some militants to 'Snap this!' and 'Snap that!' with my camera, but am then scolded by others for not moving on quickly enough. I pray they will not find out that my digital camera is broken and is registering precisely nothing.

When I need to urinate, a young man with a rifle is sent to stand next to me in the bushes. He could be my protector or my captor, although both roles seem superfluous. There is no imminent danger, and I certainly have nowhere to run to. But the gunman's presence seems to make my bladder shut down, a delay that I sense is irritating him, making it still more difficult for me to pee.

I realize that my absurd situation reveals that I am in the hands of arbitrary, unchallengeable power. To deal with it, Glenn and I are adopting an approach that is pragmatic, self-censoring and submissive. I feel as if I am beginning to have a small, controlled taste of what it means to be a poor Nigerian struggling against an abusive state. The consequences of a mis-step are potentially deadly. It is, perhaps, one of the reasons why the country has never yet risen up in the revolution that the condition of its people seems so richly to merit.

As the tour goes on, the militants' drinking continues. At the ruined and deserted local hospital, one of them trips in a hole and falls, causing his gun to clatter on the ground. The man who had shouted at us – we still don't know his name and are afraid to ask – scolds his fellow fighter and orders him to hand his gun to someone else.

The mood of our nemesis seems to darken as we process from hospital to school to an unfinished road, where an abandoned dump truck and roller have literally run into the sand. Walking down the crowded main street, he suddenly begins firing rounds in the air. There is no gasp of surprise, suggesting this is not the first time this has happened. His fellow fighters laugh a little nervously and deferentially call him 'Mr Speaker'.

As he continues to fire – five shots in total, despite Glenn's attempts to calm him down – I see a terrified-looking little girl shuffle a smaller boy protectively behind her. These children, I realize, are growing up to see guns as a normal part of life. I feel a suicidal urge to run up to the guy and ask him why he is doing this and terrifying his own people, the very community he is supposed to be fighting for. It seems to me a frightening glimpse of the civil war I'd briefly seen in Liberia, where the ideology of struggle became smothered by the seduction of the absolute power offered by the rule of the rifle.

As we leave Korokorose – not a place I am sorry to put behind me – Freeman tells me about some Korean oil workers his group kidnapped the previous year. He brought them noodles from Yenagoa, so they would not have to eat Nigerian food. They were nervous at first but then got into the swing of things, hunting, 'enjoying, sleeping, drinking'. 'We told them: feel free,' he says, as if he had invited them to a beach party.

A short way out into the river, the boat stops and we idle for a while. Freeman says we are looking for a signal from the field marshal's camp to tell us we can land. He puts on a red headscarf, replacing the white one he was wearing earlier. Asked why, he replies, less than reassuringly, 'I am going to war now.'

A few minutes later we start moving again and soon arrive at the camp where, what seems like an age ago, we had seen the naked men swimming. But instead of mooring, we turn off left towards a bend in the river. There, we put in at a jetty, and are shown to a little wooden lean-to near the shore, where a handful of fighters are sitting around or lying on mattresses. They are the watchmen, the frontline guards against incursions by the army. One of them shows us bullet marks in a tree that he says was strafed by a military patrol. 'We will not spare one soul,' he says.

After a short wait, a succession of about 30 gunshots from the direction of the camp announces the arrival by boat of the man we have been brought here to interview. Young, fit, and wearing a street-chic outfit of black T-shirt and trainers, he introduces himself simply as Olotu. He starts our interview with a long, Biblical lecture about his role in leading the fighters towards the promised land. Here

again, the contradictions flow as freely as his words: he is angry with
the government, yet he spent two years serving in the administration
of DSP Alamieyeseigha, the man at the centre of the alleged graft
case. The more he talks, the more I form the impression that his anger
is less that the status quo is corrupt, but that it displaced the equally
corrupt *status quo ante* in which he had an interest.

When Olotu gets up to leave, we swap telephone numbers and he
promises to introduce me to his sister in London. He cocks his rifle,
causing a shell to spring out and hit my boot. He retrieves it carefully
and puts it back in the magazine. As he makes for the boat, I ask
him what exactly is his role in MEND, which seems to have kept its
command structures and operations deliberately nebulous. 'Don't you
worry about it,' he grins, and is gone.

We drive back to the camp and wait a bit, until another militant
speedboat arrives. We are 'sold' some fuel that Glenn suspects the
militants stole from our boatman earlier. We settle a price and then
it is time to go: there is 'special work' going on at the camp, says
Freeman, so we cannot visit it today. A volley of shots fired from the
shore into the water sees us on our way.

On the way back, we stop again at the mobile police checkpoint we
had visited on the way out. Our interrogator is personable, scolding
me on the harshness of the British immigration system. When he
learns that Glenn and I are journalists, he says he hopes that we will
report how people are suffering in these parts. He is Ijaw and has his
own white Egbesu flag outside the checkpoint. I wonder whether he
is really opposed to the militants, or whether, like some of the fighters,
he will one day leave the security forces to take up arms against the
government. The Delta conflict's dynamics are always shifting, which
is one of the things that makes it so unpredictable and dangerous –
but also, perhaps, means ultimately that it can be resolved.

A little over a week later, back at my desk at the *Financial Times*
in London, I receive a text message that contains a disturbing final
thought on my Delta trip. As I read it, I can hear a colleague chatter-
ing on his mobile phone to an official from Shell, whose headquarters
– where I interviewed Basil Omiyi – are a little further west down

the banks of the Thames. My message is from Olotu, the commander at the MEND camp I'd visited. Carefully crafted in SMS language, it is tailored, to the character, to fit into exactly four pages. It is an appropriately frenetic and alarming postscript to my visit:

SIR, DID U HAD OF D LAST SHOOT OUT IN YENAGOA A WEEK AGO? 'A' DIVISION POLICE STATION WAS RAISD DOWN & ALL DETAINEES WERE RELEASD, SOME PART OF GOV'T HOUSE WERE ALSO BROUGHT DOWN. IT LASTD FOR 5 GOOD HOURS. OUR DEMANDS WAS DAT D PRESENT DEPUTY GOVERNOR SHLD QUIT HIS SIT 4 OTHR CONTESTANCE & NEVA 2 RECONTEST AGAIN, SINCE HE WAS RESPOSIBLE IN D REMOVAL OF ALAMIEYESEIGHA, SECONDLY WAS HIS DISAPOINT 2WORDS D NIGER DELTA STRUGGLE & OTHERWISE. WE KNW DAT DICTATORSHIP IS ONLY AN ABITRARY BORDER, WAITIN 2 BE CROSSD BY ADVNTIOUS YOUNG MEN WIT TIME. MIC DATE & U WIL KEEP RECORD.

Olotu's buccaneering, power-hungry, testosterone-fuelled text rhetoric seems to me to echo down the years, through the behaviour of the Royal Niger Company, the oil multinationals and the sequence of home-grown Nigerian autocrats whom the world either nurtured or did little to stop. It's a story whose vast scope and implications I am still trying to understand more than eight years after my first brush with Nigeria. It's also a tale that, each time I look more deeply at it, feels ever more closely linked to my own life and times.

Olotu's sign-off to his message seems a truncated cry from a country struggling to escape an abusive history, a violent present and a forbidding future. Whatever its ultimate meaning, it looks like a warning that it would be very dangerous for the world to ignore. It reads, simply:

WE HV LOST PATIENCE.

EPILOGUE:
THE HOPE OF THE WORLD

Of all the baroque encounters I've had during my years entangled with Nigerian oil, perhaps the most surreal of all was with Queen Elizabeth II. I met her at a 2003 reception held in her honour at the British High Commissioner's residence in Abuja, which was hosting the Commonwealth heads of government summit. The guests were the usual diplomatic hotch-potch, including Jack Straw, then Britain's foreign secretary, and Governor DSP Alamieyeseigha, in the days before Scotland Yard decided his wealth was worth a closer look. I had – as usual – arrived late, barrelling onto the lawn to be ushered immediately into a line of dignitaries waiting to meet the monarch.

When I was introduced to the Queen by Sir Philip Thomas, the high commissioner, I told her – a trifle disingenuously – that I'd enjoyed seeing her at her visit to a local market earlier in the day. I knew this trip had caused a little embarrassment to the diplomats, as the 'market' was actually mocked up in a local government compound, because of security concerns about exposing the royal personage to the chaos of the real thing.

It was clear that the Queen had picked up on this, too. It was fascinating, she told me, in the kind of confiding tone I imagined she might use in her weekly meetings with the prime minister, although some people had told her it wasn't the *proper* Nigeria'.

The high commissioner – who was to become a consultant for Shell after he retired – quickly applied some foreign office balm to her concerns. 'It was a *blend*,' he said.

Then the Queen asked what I wrote about in Nigeria. Well, I said, there was a lot of oil there so I spent quite a bit of time examining that.

Her Majesty looked at me with a gaze that suddenly seemed more acute, intense and – I fancied – mischievous.

Then, softly, she said, 'It doesn't always *help*.'

More than half a decade on, the monarch's observation is being echoed loudly by many Nigerians and outsiders who know the place a bit.

A few years after the Queen's visit, the 2007 presidential elections that were supposed to mark Nigeria's first transition from one civilian president to another turned into a fiasco. On polling day, I drove around Port Harcourt and its hinterland and saw the same blanket fraud and intimidation that I'd witnessed four years earlier, in the first election after the end of dictatorship. Late in the afternoon, some Nigerian observers and I followed up on a tip that ballot boxes had been taken to the house of a powerful local politician. We drove towards it, only to be stopped by a roadblock made of logs and superintended by a group of young men. We could go no further, one of them explained politely, as there was a 'communal war' going on, and they did not want us to become tangled up in it. We had reached the end of our road.

Western nations joined a chorus of Nigerians in criticizing the elections, but recognized the new government of President Umaru Yar'Adua nonetheless. The new leader's first 21 months inspired little confidence. High-profile officials identified with reform efforts either slipped from the scene or were frozen out. Nuhu Ribadu, head of the Economic and Financial Crimes Commission, was sent on a mysterious study leave shortly after he had arrested James Ibori, a powerful former oil-state governor with close connections to the president's People's Democratic Party. Later, Ribadu was dismissed from the police force – to protests from anti-corruption campaigners – for alleged 'insubordination'. Ribadu and his crime-busting commission had a mixed record, but many Nigerians agreed they had been one of the few forces in the country that had provided any kind of check on graft.

President Yar'Adua himself seemed a weak and perpetually diminishing figure, his health a constant subject of speculation as he travelled abroad for medical treatment. The steep drop in the oil markets amid looming recession in leading economies – the price plunged from $132.55 a barrel in July 2008 to $40.76 in December – promised further problems for the leader and his restless country.

The sense of political drift was captured by a dark popular joke about the renaming of the notoriously unreliable National Electric Power Authority as the Power Holding Company of Nigeria (PLC). As the initials were switched, so did what they were said to stand for: 'Never Expect Power Anytime' became 'Problem Has Changed Name (Please Light Candle)'.

Home-grown plutocrats like the billionaire Aliko Dangote flourished, along with other merely well-off members of the sizeable business class, but there were scant signs of relief for the masses.

A little over a year into the new administration, a lawyer friend of mine not given to histrionics wrote me a desperate e-mail begging the Western media to draw attention to a country that was 'fast slipping under the firm control of elements intent on ensuring that the people remain mired in poverty and decay'. 'If Nigeria becomes openly and clearly under the control of obvious criminals,' he lamented, 'the damage would be far greater than anything else that's going on in Africa.'

In February 2009, MEND dramatically called off a Delta ceasefire it had declared five months previously. It kidnapped the wife of Edmund Daukoru, former energy minister, and attacked a gas plant. MEND said its action was meant 'to send a message to the oil companies that all the pipelines they have repaired in the Western Delta will soon be in need of repairs again'.

Weeks later information minister Dora Akunyili – much respected for her work in a previous job combating the fake drugs that plagued street markets – launched Nigeria's latest attempt to rebrand itself. Unveiling the slogan, 'Nigeria: Good people, great country', Akunyili said the nation needed to shed its image as 'untrustworthy, unreliable and ungovernable'. The impulse behind the initiative was understandable, to a point: there is – as the author Chinua Achebe once put it and some prejudiced foreigners still need reminding – 'nothing basically wrong with the Nigerian character'. But the danger with such sloganizing, as with all nationalistic PR stunts, is that they can be used by less scrupulous officials than Akunyili as an excuse for dismissing legitimate criticism, both domestically and overseas.

In May 2009, another spectre from Nigeria's unresolved past was due to rise, in a US court case in which Shell was accused of complicity in the 1995 execution of Ken Saro-Wiwa. The company dismissed the claims – brought by a group that included Ken Saro-Wiwa Jr, the late activist's eldest son – as false.

For all the limited improvements since the end of dictatorship in 1999, Nigeria looked as if it was treading ever further down a path that would prove very dangerous for both itself and the rest of an oil-dependent world. When I'd first visited the country six months before the 11 September 2001 terrorist attacks, it had seemed to be seen internationally as a middling crude producer of middling global importance with middling problems. More than seven years later, as the Middle East simmers, the Niger Delta seethes and the Chinese and Russians arrive, Nigeria looks both more strategically important and more dangerous. As one of the losing Nigerian presidential candidates said to me shortly after the 2007 elections, it seemed a country that was sure to blow, even if no one yet knew quite when or how. For all the clear and acknowledged problems, no one seemed to have much of a plan beyond narrow – and morally questionable – investments in tougher protection for existing oil and gas installations, and in bringing new production onstream.

The cocktail of hostage-taking and election-rigging I'd seen in the Delta showed how power was continuing to shift towards thugs and militias of all stripes, inside and outside the structures of the state. Approaching ten years of civilian rule, the cult of corruption, secrecy and violence that had grown up around the country's oil industry remained entrenched. I thought, not for the first time, that the only way out might be to have a kind of South Africa-style truth and reconciliation commission over oil. Then all sides – the government, the oil companies, the gangster-militants and the Western consumers – could admit the bad things in which they were complicit, change their behaviour, and move on.

In the years since my first involvement with Nigeria, I realized I had come to look at what has happened to the country not as *sui generis*, but as an emblem of a parasitic and potentially cataclysmic era of human history dedicated to short-term material gain. The more I travelled between West Africa and Europe, the more I was struck not by the differences, but by the parallels of poverty among riches, institutional cruelty and societies in which élites talked to and of themselves. Oil may have made the Nigerian canvas particularly vivid, but the portrait it depicted was instantly recognizable. Nigeria offers us a terrifying vision of the consequences for us all of tolerating inequality, profligate energy use and environmental disaster.

The Pandora's box hope for Nigeria, an independent nation for less than 50 years, is that – if it is ever given the chance – it still has all the opportunities to change character that youth allows. There is no polish or veneer of order to this vast, messy country, but nor is there the same sense of secret histories and hidden hypocrisies that lie behind the swept streets, ancient institutions and grand public buildings of Britain and other rich nations. In oil-driven Nigeria, the exploitation, injustices and abuses of power are more open, more blatant and, in a strange way, more honest. That should make radical change more possible than in a country such as Britain, where injustices have become more entrenched and subtly concealed over many centuries. As the West declines in relative importance as a centre of political and economic power, so the opportunity is there for the peoples of other places to offer new ways of thinking and a hunger for reform. Nowhere else, among the countries I've seen, is this appetite for change – and the creativity to deliver it – greater than in Nigeria, where the axiom that no condition is permanent is applied with a brio unmatched by most Western workers going about their daily grind.

It is a measure of Nigerians' extraordinary vigour that they continue to drive the country on despite all the problems the age of oil has bought. Ground down but defiant, many people in this nation of huge, dissipated energy and intellect seemed to have mastered the trick – the author Rohinton Mistry's 'fine balance' – of combining deep short-term pessimism with a peculiar longer-term optimism. Again, just like Tennyson's Ulysses, many Nigerians continue to strive, to seek, to find and not to yield. On one visit to Lagos, I even saw a banner near the airport that proclaimed the nation 'the hope of the world'.

As I prepared to leave Lagos, not knowing when I might come back, I had one more glimpse of the latent spirit that could one day help Nigeria – and, if it wants to listen to these lessons from West Africa, the wider world – throw off the dead hand of oil dependence and find its way to a better future. I was stuck on a motorbike taxi on the water-side, just down from a dishevelled old ship hulk decorated with the ominous warning to 'keep Lagos clean or leave'. Everyone was trying to make the right turn to Victoria Island, although the route straight on that I needed was empty. I was reminded of how often in Nigeria I had the frustration of being boxed in by cars yet able to see clear road ahead.

As we nudged through, a security van pushed its way past us, its siren whining like a dirge for the trapped commuters all around. I started talking to my driver, Fabian, about the problems of living in Nigeria. He told me his mother and father had died, so he had come to Lagos to find work. He needed help, because motorbike taxi driving was not a proper job. Like many Nigerians, he was filling time industriously until something better came along.

After a few minutes of skilful edging, ducking and weaving, Fabian finally got us to the point where the road divided. A well-timed burst of gas took us clear, flying off the curve of traffic at a glorious tangent. We both started laughing and whooping at this small victory over circumstance, another unpromising moment seized and squeezed for every small advantage it contained. 'Yeah, yeah, we are free!' Fabian shouted, as we careered up the slip road, soaring above the gridlock that would probably paralyse the shoreline of Africa's oil giant until nightfall and beyond.

Michael Peel, April 2009

NOTES AND SELECT BIBLIOGRAPHY

This book is based on research trips to West Africa in 2005, 2007 and 2008, building on my experiences as the Lagos-based correspondent of the *Financial Times* between 2002 and 2005. Where I think it is relevant, I have been specific about the dates of meetings and visits. I've tried to bring everything as up to date as possible as of going to press. Any failures to do so – like any other errors of fact or interpretation – are entirely my responsibility.

For statistics, I have tried to choose forms of data that are easiest to understand and most comparable. The international oil prices are weighted International Monetary Fund monthly averages. Historic exchange rates are those used at the time of the events under discussion.

Where there are disagreements over spellings of place names or of words in pidgin English, I have tried to use the most commonly employed form. In a few cases, I have amended Nigerian English grammar to make it more easily understandable to a British English reader. In one case – my visit to the Lagos buses – I have changed the name of an interviewee, 'Chiguzor', to protect his identity.

I am indebted to a range of books on Nigeria, from internationally acclaimed fiction to the many detailed non-fiction works that are unlikely to be found in many places outside Lagos's *Glendora* bookshop. Some authors – such as Chinua Achebe, Wole Soyinka and Chimamanda Ngozi Adichie – are figures of world stature who need no introduction from me. Other writers well worth looking up if you don't know them already include Adewale Maja-Pearce, Okey Ndibe, Dulue Mbachu, Elechi Amadi and Buchi Emecheta.

Among non-fiction, *Where Vultures Feast* by Oronto Douglas and Ike Okonta is a passionate and insightful account of the crisis in the Niger Delta. *The Next Gulf* by Andrew Rowell, James Marriott and Lorne Stockman takes a broad and helpful look at Nigeria and the politics of world oil. I found both Michael Crowder's *A History of West Africa Under Colonial Rule* and the British National Archives records very useful in understanding Nigeria's place in the world imperialist jigsaw.

Finally, Omoyele Sowore and his colleagues at Sahara Reporters – *www.saharareporters.com* – have delivered regular brilliant exposés on the vast nexus of corruption around oil in Nigeria.

ACKNOWLEDGEMENTS

Two pages are scant space to pay tribute to the many people who have given me companionship, ideas and material during the eight years of my association with Nigeria. Many – including one in particular – will have to remain nameless, but I've hugely enjoyed the friendship and insights of C. Don Adinuba, Desmond Majekodunmi, Kaye Whiteman, Okey Ndibe, Tunji Lardner, Phil Hall, Bill Knight, Tola Olateru-Olagbegi and Chinyelu Onwurah.

I have been privileged to enjoy the professional brilliance and splendid company of many members of Lagos's small foreign press corps, notably Daniel Balint-Kurti, Anna Borzello, David Clark, D'Arcy Doran, Glenn McKenzie, Dino Mahtani, Dan Isaacs and Jacob Silberberg. These journalists, who have worked to bring a little bit of the Nigerian story to a Western audience each day, are among the very best I have come across during my career. I am very grateful for the contacts and thoughts they provided during some of the visits described in this book, as well as on many other occasions.

Of the many people who helped me stay well and navigate Nigeria's often fearsome but endlessly stimulating logistical challenges, I owe a particular debt to Gabriel Akinyemi, Gabriel Bawa, Rasaki Hassan, Maurice Okon and Ayuba Martins, may he rest in peace.

Socially, I had some of the sweetest moments of my life in Nigeria. In addition to many of those named above, Vera Ephraim and Inna Erizia deserve special thanks for that.

The BP press fellowship programme at Wolfson College, Cambridge, provided a welcome breathing space to work on this book and to enjoy a nostalgic return to student life. I am grateful to John Naughton, Bill Kirkman, Hilary Pennington and, most of all, to Richard Synge, who gave me many ideas and generous access to both his time and his rich archive of material on Nigeria during the 1970s and 1980s.

Another source of support was my associate fellowship at Chatham House, where I owe great thanks to Alex Vines and Tom Cargill.

Camilla Watson and Kathleen Becker offered me wonderful hospitality and contacts for my trip to São Tomé and Príncipe.

I deeply appreciate the many people who read and commented on this book in part or in whole, some of them probably more convinced than I was that it would see the light of day. They include Lara Pawson, Julian Richards, Chimamanda Ngozi Adichie, Lawrence Booth and Michela Wrong. Thomas Penn was a great touchstone and source of encouragement, as he has been for the past quarter-century.

Many *FT* colleagues past and present helped with the production of this book. As for so many other young journalists with an interest in the African continent, Michael Holman was an inspirational mentor. My thanks go also to Lionel Barber, Dan Bogler, Nick Timmins, David White, Matthew Green, Antony Goldman and William Wallis. Megan Murphy has been a wonderful colleague, friend and adviser.

At I.B.Tauris, I am very grateful to Abigail Fielding-Smith, who breathed life into this project just when I feared it was about to die. Thanks also to Jayne Hill at I.B.Tauris and Carolann Martin at Initial Typesetting Services, who finally tore the manuscript away from me and made sure it was published. Georgina Palffy was a fantastic close editor and Linda McQueen did a great job on the proofing.

Finally, two big cheers. One is to all the family and friends who have given me support and – just as importantly – graciously tolerated my absences and distractedness during the production of this book. Most of all, I would like to thank my parents, Mary and Robin Peel, and Samantha Lister, for their great love and forbearance.

The other cheer is to the legion of people in Nigeria and elsewhere in West Africa who have shared something of their lives with me over the years. Those times – and my gratitude for them – will be with me always.

INDEX